INTERNATIONAL WILDFOWL INQUIRY

VOLUME I

FACTORS AFFECTING THE GENERAL STATUS OF WILD GEESE AND WILD DUCK

CAMBRIDGE
At the University Press
1941

CAMBRIDGE UNIVERSITY PRESS
Cambridge, New York, Melbourne, Madrid, Cape Town,
Singapore, São Paulo, Delhi, Tokyo, Mexico City

Cambridge University Press
The Edinburgh Building, Cambridge CB2 8RU, UK

Published in the United States of America by Cambridge University Press, New York

www.cambridge.org
Information on this title: www.cambridge.org/9781107617483

First published 1941
First paperback edition 2011

A catalogue record for this publication is available from the British Library

ISBN 978-1-107-61748-3 Paperback

Uniform with this volume

J. BERRY. The Status and Distribution of Wild Geese and Wild Duck in Scotland

(*International Wildfowl Inquiry*, Vol. II)

THE INTERNATIONAL COMMITTEE FOR BIRD PRESERVATION

(*BRITISH SECTION*)

WILDFOWL INQUIRY SUB-COMMITTEE

CHAIRMAN
Percy R. Lowe, O.B.E.

Dr A. Landsborough Thomson, C.B. (Royal Society of Edinburgh)
David Seth-Smith (Zoological Society of London)
Dr G. Carmichael Low (British Ornithologists' Union)
N. B. Kinnear (National Trust)
Hugh Whistler (National Trust)
G. E. Lodge (Society for the Promotion of Nature Reserves)
J. C. F. Fryer (Ministry of Agriculture and Fisheries)
R. A. H. Coombes (British Trust for Ornithology)
Stanley Duncan (Wildfowlers' Association of Great Britain and Ireland)
J. A. S. Stendall (Ulster Society for the Protection of Birds and *Irish Naturalists Journal*)
G. R. Humphreys (Ornithologists in Eire)
T. Russell Goddard (Hancock Museum, Newcastle-on-Tyne)
Eric Parker (formerly Editor-in-Chief of the *Field*)
H. Calkin (*Country Life*)
Major Hugh Pollard (Shooting Editor, *Country Life*)
Major G. Burrard, D.S.O. (Editor of *Game and Gun*)
Noel M. Sedgwick (Editor of the *Shooting Times*)
Miss E. P. Leach (Hon. Sec. Bird Ringing Committee, British Trust for Ornithology)
Dr John Berry
C. W. Chaworth-Musters
Major H. P. S. Clogstoun
C. T. Dalgety
H. A. Gilbert
The Earl of Ilchester
J. C. Laidley
Colin McLean
C. W. Mackworth Praed
H. Leyborne Popham
J. W. Stares
H. Tetley
C. W. B. Urmston
Jim Vincent

SECRETARY

Miss Phyllis Barclay-Smith, c/o The Zoological Society of London, Regent's Park, London, N.W. 8

CONTENTS

FOREWORD

For two or three years the attention of a special Sub-Committee of the British Section of the International Committee for Bird Preservation has been concentrated on an inquiry into the present status of wildfowl (wild duck and geese) in the British Isles and elsewhere. Concurrently with this inquiry other European nations had agreed, at a Conference of the International Committee held in Vienna in 1937, to institute similar investigations in their own countries. These inquiries, the importance of which in so far as the future welfare of wildfowl is concerned can hardly be overestimated, were primarily suggested by the very unsatisfactory state of affairs which prevailed in the United States of America. Preliminary investigations had revealed that a similar condition of things might very soon be expected in Europe, and indeed throughout the world, unless steps were taken in good time to counter the various adverse factors which were combining to render this result almost a certainty.

It was, for instance, quickly realised that in less than a hundred years a complete revolution had occurred in the combination of factors which controlled the numerical status of the world's stock of wildfowl.

It was also realised that unless the nations concerned got together in a firm determination to counteract these factors, in so far as they could be counteracted; unless, that is to say, there was a definite attempt towards reciprocal action among nations and a liberal give and take attitude in their policy regarding the necessary measures to be taken, the outlook for the future welfare of the wildfowl of Europe was very gloomy indeed. In this respect both the economic and sporting value of the wildfowl was bound to be affected. It was therefore of the first importance to discover the nature of the factors which in so many different ways and directions were exerting such a baneful influence on the world's stock of Anatidae.

In this introductory volume of the Committee's report some of these adverse factors have been considered in a general way and some in a more particular way, as the result of personal experiences and contacts on the spot.

The actual status of British wildfowl will be considered in subsequent volumes (and it is hoped, also, in due course to publish similar accounts of the status of wildfowl in other countries of Europe).

It will be obvious, however, that the status of wildfowl in any country may be directly affected by factors which may have their seat of origin far afield in northern as well as other latitudes. In gaining a knowledge of some of these factors the Committee may be considered to have been very fortunate in having secured the assistance of certain well-known field ornithologists and sportsmen who have been accustomed to visit northern latitudes from year to year and who have in consequence acquired a practical and first-hand knowledge of local conditions. In many cases the knowledge which they possessed of the actual state of affairs in the north goes to the very root of the trouble. One has only, for instance, to read the reports of Major W. M. Congreve, Captain McNeile, Dr Steward, Messrs Dalgety and Freme and the brothers Bird, on the conditions which obtain in Iceland, Greenland, Spitsbergen and elsewhere, to realise how very valuable is their testimony. We know of no reports of the kind which have been previously published and it only requires equally authenticated notes and warnings on the part of other nations to arouse a proper attitude through the whole of Europe to the real danger which threatens our wildfowl.

In the same way the report of Dr R. W. Butcher, of the Ministry of Agriculture and Fisheries Research Laboratory, Nottingham, on the decrease and present distribution of *Zostera* calls for our very warm praise. It will be read by all wild-fowlers, not to mention scientific botanists, with the greatest interest and profit. *Zostera*, as most of us know, has been considered to play a very important part in the food consumption of the Brent Goose and Wigeon. Dr Butcher has made a very special study of it and kindred marine plants, and in this volume gives us the full and liberal benefit of his investigations. We venture to say that no more valuable, comprehensive and practically useful report has ever been written.

Then too, from the pen of Captain H. A. Gilbert, assisted in this compilation by Messrs Mackworth Praed, Colin McLean and H. Leyborne Popham, we have a concise history and epitome of the present status of duck decoys. Too many wildfowlers are inclined to think that as regards their sport all is for the best in the best of all possible worlds. It is obvious that they take too parochial a view of things. "Let well alone", they say, "there are just as many duck nowadays as ever there were." They look back perhaps twenty or perhaps thirty years and possibly no further afield than their own county. They are living, we suggest, in a fool's paradise. But Captain Gilbert's facts about decoys ought to convince them that if "there were as many duck as ever there were" we should have as many decoys. In the Report on Punt-gunning Mr C. T. Dalgety, assisted by Messrs R. A. H. Coombes, H. A. Gilbert, H. Leyborne Popham and T. Russell Goddard, pours some cold truth upon continental allegations that British punt-gunners are in the habit of slaughtering hecatombs of duck. Whatever may be thought of the ethics of punt-gunning, the actual practice and technique of getting within shot cannot be considered in any other light than as a sporting event, and readers will probably be much surprised at the small numbers of shots and the small average yearly bags obtained by such comparatively few punts as are employed.

The Ringing of Wild Duck is another report of very considerable value which has been undertaken by Dr A. Landsborough Thomson. As a now well-established means of tracing the migratory paths and spring or autumn objectives of wild duck, this comparatively new branch of study needs no introduction to either wildfowlers or ornithologists. Dr Landsborough Thomson has brought together a mass of instructive notes which will be read with great interest and benefit by all. In his illuminating account of the ringing being done at Orielton Decoy, Mr Mackworth Praed proves the inestimable value of this means of ringing duck and the many records thus obtained. No less interesting are the results of similar work being carried on at his decoy at Abbotsbury by the Earl of Ilchester and at Borough Fen and Somerleyton by Mr Colin McLean.

Finally we have an expatiation by Miss Phyllis Barclay-Smith on Close Time, on former wild-bird Acts in Great Britain, and on the present existing Wild Birds (Duck and Geese) Protection Act, promoted by the British Section of the International Committee for Bird Preservation. Like many other good Bills presented to our legislative chambers, the Act which finally emerged with its wings clipped by the apparently inevitable compromises was not the "perfect Act" which the Committee had envisaged. It was not an easy task to explain its inhibitions to the man in the street: in this we venture to think Miss Barclay-Smith has succeeded admirably.

In conclusion, I must place on record our appreciative thanks to the many sportsmen and naturalists all over the country who have given us their whole-hearted assistance by their invaluable personal first-hand observations, and views; to Dr R. W. Butcher and Miss Strudwick for allowing the use of their illustrations on *Zostera* from their work *Further Illustrations of British Plants*, and to the publishers of that book, Messrs L. Reeve, Ashford, Kent, and to the

Editor of the *Journal du Conseil International pour l'Exploration de la Mer* for permission to use the illustrations from Dr Butcher's paper "*Zostera*: Report on the Present Condition of Eel Grass on the Coasts of England", published in that journal. To the generosity of the British Ornithologists Union, the British Ornithologists Club, the Royal Society for the Protection of Birds, individual subscribers and the Syndics of the Cambridge University Press, whose financial aid has made the publication of this volume possible, we owe a great debt of gratitude.

PERCY R. LOWE

Chairman European Continental Section
Chairman British National Section
International Committee for Bird Preservation

LONDON 1940

I

INTRODUCTION

THE HISTORY OF EVENTS LEADING TO THE FORMATION OF THE WILDFOWL INQUIRY SUB-COMMITTEE

BY PERCY R. LOWE (CHAIRMAN)

It will probably be admitted by anyone who has given serious thought to the question, that during the last fifty years or so there has been a steady diminution in the world's stock of Anatidae in general. This diminution has lately become so obvious in certain quarters as to give rise not only to local but to general concern.

The first indication that all was not well with the world's stock of Anatidae was, I think, made evident by the constant bickerings which took place on the subject of the common stock shared between Canada and the United States of America. Canada, for example, was constantly complaining that she bred the duck and geese and the United States shot them. She complained, moreover, that the United States shot them thoughtlessly and without any regard to their future welfare as a whole, without any regard to Canada's interests or to her own.

I mention these little bickerings because in Europe we have a very similar state of affairs; that is to say, we have in general terms the producer in the north and the profiteer further south. Under such conditions it is clear that unless there is some general convention of a give-and-take nature, we shall in Europe sooner or later be faced with the same serious problem which is now facing the people of the New World.

It may be laid down as a general principle that in a community of nations like Europe, no individual nation has any *absolute* property right in the migratory duck or geese which come from the producing countries in the north to winter within the limits of her territory, or which traverse it as passage migrants in the spring and autumn. No nation has any moral right to shoot these duck without limit, regardless of the interest of other nations, or, in the long run, of her own. She receives them on the basis, so to speak, of a sort of entail. It is every nation's bounden duty to see that this entail is not broken. It has to all intents been broken in one notorious instance on the continent. That is to say, the wild duck which used to winter in the country referred to have long ago been practically exterminated by ill-regulated shooting. As a consequence not only has that particular link in the migratory chain concerned been broken, leading to repercussions in other countries, but her sportsmen are now reduced to an organised shooting of passage migrants, which pass along her shores in the late spring. These migrants are already paired and on their way north to breed. At this late spring season the ovaries of the female duck are full of potential young. Need we wonder that the producing countries in the north where these duck nest have a just cause of complaint. But the country to which we have referred is not the only one to blame; for there are cliques of so-called sportsmen in the British Isles who wish to shoot Wigeon at just such a late date, and others in Scotland who do not even pretend to be sportsmen, who would shoot wild geese later still.

In the New World attempts have been made by Canada and the United States to observe some such principles of reciprocity, to which we have referred, by conventions, and unless measures of a like nature are taken in Europe the outlook for the future status of the European Anatidae is very disquieting. These measures in the New World have apparently not been adopted a moment too soon. In a pamphlet with a foreword by Mr Kermit Roosevelt, President of the National Association of Audubon Societies, and published* in 1937, he states that as a result of investigations undertaken by three independent inquirers in three different parts of the United States, the conditions disclosed in regard to the shooting of water-fowl were, to use his own words, "profoundly shocking". Mr Kermit Roosevelt's reputation as a good sportsman is not such as to lay him open to a charge of exaggeration or sentimentalism, and one of the above-mentioned investigators even goes so far as to say in his report, "I am not a sentimentalist and have probably sold as much duck ammunition and shot guns as any one person in America." Conditions indeed are so bad that the view is expressed that it is possible that all shooting of wildfowl will have to be stopped as a temporary measure to restore the balance between supply and demand. In point of fact we understand that it has been limited to what amounts to one month only in the year for each state.

* *Ducks of To-morrow*, National Association of Audubon Societies. New York, 1775 Broadway, 1937.

Passing over from the New to the Old World we find that Professor Einar Lönnberg of Sweden was the first to bring the subject of the diminution in the numerical status of wildfowl to the notice of European ornithologists and preservationists. This was in 1925. Lönnberg's efforts were followed up in 1926 by the Swedish Government itself in the form of official proposals to European Governments which had for their object the establishment of international regulations aimed at the more effective protection of wildfowl on migration.

In the same year, at the International Ornithological Congress held in Copenhagen, I brought forward a resolution suggesting that reports on the subject should be drawn up by representatives of the various nations concerned, and that an International Conference should be convened to consider the question.

This conference was duly held in the Foreign Office in London at the invitation of H.M. Government and under the Chairmanship of Lord Ullswater. As a result a report was published by the Home Office in which certain recommendations were embodied, notably the shortening of the open season and the investigation of migratory routes by means of ringing. On the whole, however, the results of the conference were disappointing, due in large part to opposition based on vested interests connected with decoying.

There followed years of comparative lethargy and then, in 1936, there came the startling news, referred to above, that the United States of America had practically limited the shooting of wild duck to one month in the year. This, apparently, was a temporary measure, but it pointed to a grave state of affairs. About this time (1934) Mr J. N. Darling, Chief of the United States Biological Survey, had estimated that only from 18 to 20 million duck were left on the North American continent. It may be granted that these are rough figures, but whatever the exact figures might have been, no less than 600,000 licences to shoot these duck were taken out in 1934 which, as he points out, leaves less than forty birds for every licensed shooter. To those who can remember seeing the uncountable hordes of North American duck wintering in Mexico and other regions, these figures point to a condition of things which can only symbolise the beginning of the end. Thirty-two years ago I have myself shot duck on the east coast of Mexico at such a rate that the barrels of my gun became literally too hot to hold; and not on one occasion only or in one year. I mention the fact as a concrete example within my own memory of the phenomenal diminution in the numerical status of North American duck.

Coming back again to the Old World we now hear that Sweden has provisionally stopped all shooting of duck in Jamtland, one of her largest provinces, for five years; and has taken similar measures in respect of one or two other provinces. Such an action can only mean one thing, too obvious to dwell upon. Nearer home we have the melancholy spectacle of France practically "shot out", as regards her wild duck, and apparently politically powerless to remedy such a state of affairs. We learn too that, in 1935, 1,450,000 licences to shoot birds in general were issued in that country.

In Holland it has been estimated that something approaching 1,000,000 (one million) duck are annually taken in decoys, a share which in view of the present European stock is by general consent considered disproportionate. The Decoy Group, in their very complete and instructive report on the Decoys of Great Britain and Europe (see page 50), has considered this side of the problem, and has, so to speak, added another chapter to Payne-Gallwey's now classical history of the decoys of Great Britain, in which he shows that, as far as the latter country is concerned, decoying is a nearly dead or negligible industry.

There are some wildfowlers who take a narrow or even parochial view of the present status of wild duck in this country, and who go so far as to say "there are just as many duck as ever there were". If there is any one thing more than another which could be quoted as proving the absurdity of such a contention so far as migratory and not home-bred duck are concerned, it is the history of decoying in England (see Decoy Report) during the last eighty years. A hundred years ago the skies over such marshy areas as the fens used to be literally black with dense masses of duck, either on migration or moving from one district to another. Decoying was then a really paying business.

In the marshland of Hortobagy in Hungary we also learn that it is the ambition of sportsmen to shoot hundreds of geese in one day. In Egypt, too, we hear of deliberate measures taken in high quarters to commercialise the actual shooting of wild duck, a proceeding which must surely be inevitably fatal to the ultimate status of duck along that line of migration, to say nothing of the project as a commercial undertaking.

Disquieting news comes too from India, where rumours of a steady decline in the numerical status of the Anatidae are persistent. This may be due to shooting on a more general scale following upon altered social and political conditions, to increased facilities of transport by means of motor-cars, or to adverse factors present on the breeding grounds in Siberia.

Thus on the world's three great migratory wildfowl tracts we have, as indicated in this brief sketch, the same definitely disquieting rumours. It is clear that the conditions which ruled in the past no longer obtain. A formidable series of factors has arisen which has completely upset the old-established balance of supply and demand. The supply from the north is less and likely to be still further reduced while the demand in the south is greater. It is clear then that we must take steps to readjust our actions in face of this revolutionary change. *What, therefore, are the factors to which we have alluded?*

To the investigation of this problem, a Sub-Committee of the British Section of the International Committee for Bird Preservation, consisting of some thirty or more members, all experts in some one or more of the various branches of inquiry involved, has given for the last few years its voluntary and invaluable aid. In the following remarks I have given a review of some of these factors which in one way or another are combining to reduce the numerical status of the European Anatidae, and we may state in passing that, although the Sub-Committee was primarily set up to investigate conditions chiefly obtaining in the British Isles, these factors have a range far beyond our territorial limits; and, as will be obvious, apply equally to all other nations. It is, therefore, of the first importance that these and other factors should be realised by those nations, and that they should come together in a strenuous endeavour to do what is possible to counteract them.

Some Factors Responsible for a Revolutionary Diminution in the World's Stock of Wildfowl

(1) *Increased Facilities of Travel and Transport*

The introduction of the steam engine, railways, steamships and the internal combustion engine may perhaps be quoted as the chief and primary factors which have transcended all others in diminishing the numerical status of wildfowl. They have revolutionised their present status. Even with the most enlightened management it is extremely unlikely that the numerical status obtaining a hundred years ago will ever again be attained.

It seems hardly necessary to mention that the factors indicated above have led to enormously increased facilities for travel and transport and these have led in turn to the opening up, the peopling and exploitation of the resources of countries previously inaccessible or only accessible with great difficulty; while in numerous instances those countries represented the previously undisturbed breeding haunts or wintering quarters of countless millions of wildfowl.

The internal combustion engine has made speedy and easy travel possible by motor-car, motor-bicycle, motor-bus, lorry, and so on. Roads have been built on which such engines of transport can penetrate smoothly, not only to every quarter of the already civilised globe, but to regions well within the Arctic Circle and the high latitudes of such countries as Scandinavia, Finland and Siberia, where wild duck and geese previously nested in undisturbed peace. In our own country modern means of transport permit access to the coast from inland districts even as far inland as Birmingham at a speed which, compared with old times, is little short of miraculous. As a consequence, week-end visits, or even day trips to our estuaries can now be made by scores in search of wildfowl shooting where formerly only those who lived in the vicinity, or a privileged and leisured few, had any chance or hope of enjoying sport. The disturbance of wildfowl is, therefore, far greater and more frequent than of old. Many more shots are fired, and fired moreover by incompetent and inexperienced people when there is no hope of killing, with the result that many birds are wounded or finally driven away to quieter quarters. The same easy and speedy means of transport gives access for hundreds of thousands of holiday seekers or week-enders to almost every mile of our coasts and, as a consequence, maritime towns are spreading in an almost continuous chain along our shores, and bungalow populations are springing up like mushrooms in every direction to usurp areas formerly claimed by Nature alone. It is now possible for a wildfowler to fly by aeroplane to the marshes of Hungary and get his quota of geese in as short a space of time as would have been consumed in the old days by travelling by rail from London to the Solway. Motor-cars are used in Egypt to follow up wild duck, driven from lagoon to lagoon, which previously would have been out of reach in any given day's shooting; and fast motor-boats are, we are informed, used in the same way on the Nile. It is now possible to travel in Finland by motor-buses to well within the tundral and Arctic areas where one can find comfortable rest-houses and inns.

(2) *Cold-storage and Commercialisation*

I should be inclined to put cold-storage as the next most deadly factor threatening the future of wildfowl. It has not only made possible the importation from foreign countries of beef and mutton and every possible variety of food, but even makes the importation of game birds and wildfowl a profitable concern. *Wildfowl are, therefore, being commercialised* in Europe wherever opportunity offers.

It has been proved over and over again that commercialisation of any *wild* species without the most stringent precautions leads almost inevitably to its extinction, or to such a dangerous depletion in numbers as to threaten its extinction. In the United States of America no wild duck or wild geese are permitted to be exposed for sale. It is a debatable question if their exposure for sale should be permitted in Europe. At any rate it may be taken as almost axiomatic that, if it were made illegal, the future outlook for wild duck and wild geese would give cause for far less anxiety. For instance, decoying duck for the market would receive its death-blow, as also would the importation of frozen wildfowl into European countries. They were even imported into this country *during the close season,* a condition of things which can only be described as courting disaster. This has now, however, been made illegal by the Act introduced by the British Section of the International Committee for Bird Preservation. In any case their importation is only rendered possible because of the existence of two factors—steam and cold-storage.

(3) *Conditions in the Far North*

We have also collected from the actual experiences of members of our committee information concerning the depredations committed by hunters in Spitzbergen and Greenland; and also in Iceland by farmers. In Spitzbergen, outlying islands, where geese and wild duck nest, are systematically plundered for their eggs without the slightest forethought for the future, and on a completely wholesale scale. In Iceland no restriction seems to be placed on the taking of the eggs of the Pinkfooted and Whitefronted Geese, to say nothing of wild duck, or if laws are passed they are not enforced. It is true that this plundering of eggs is done to provide food for the farmer and peasant during the winter, and moreover, has been done for generations without any conspicuous ill-effects in the past; but the ever-increasing toll taken by more southern European countries during the migrations of these geese and duck has introduced a factor which can shortly be described by saying that the demand is exceeding the supply. Samoyedes too, have for generations been compelled, for their very existence, to kill moulting and, therefore, temporarily flightless geese. As they have done this for generations the fact in itself can hardly be considered as a factor affecting the diminution of the numerical status of geese at the present day; but if modern facilities of transport and cold-storage have made it profitable to kill and export wild duck and wild geese or their eggs, such commercialisation must have a very much more serious effect, since not only would operations be conducted on a much larger scale, but it would be done at the very time when so many other artificial or unnatural factors are working against the numerical status of wildfowl.

(4) *Ill-considered Reclamation of Unsuitable Areas of Land*

The drainage of marsh lands, swamps and fens is in many cases not only a factor which obviously destroys the breeding haunts and wintering quarters of wildfowl and countless forms of life, but in only too many proved cases it has for its disastrous results the altering of the level of the general water-table in the country concerned, thus leading to a wholesale desiccation of large areas of contingent crop-bearing land and the resultant sweeping away by winds of the superficial soil. In the United States of America it has now been realised what a disastrous chain of events may follow these hasty and ill-considered schemes of drainage which have ruined thousands of square miles of valuable crop-bearing land. Nevertheless we are informed that unscientific and, in the end, profitless operations of a like kind have been, and are being, carried out or projected in some central European countries. It is certain that such marshes and swamps as exist in such countries play a beneficial part and act like vast sponges holding up the water and controlling the water-level in huge areas of crop-bearing land. They act in the same way as forests, the ill-considered and ignorant cutting down or destruction of which has an equally disastrous effect on contingent land areas.

(5) *Other Agricultural Factors*

Closely connected with such factors as the above is the adverse effect of the cessation of the irrigation of water meadows in certain districts, or the unwise policy of allowing water to run rapidly off the land before it has done its work as would happen under more natural conditions. In referring to agricultural factors which bear on the welfare of wildfowl, we may also draw attention to changes in cropping such as the comparatively new industry of potato farming, and of course the far greater and more serious threat to the future of wildfowl which the spread of towns in Russian Siberia or agricultural schemes in the tundral areas of Finland, Norway and Sweden represent. Former villages are becoming towns along the northern coast-line of Russian Siberia; factories are being built and areas exploited for one purpose or another to which formerly only the most adventurous travellers penetrated. All these activities cannot fail, sooner or later, to have their repercussions on the breeding grounds of the teeming populations of wildfowl which formerly existed there. Moreover, we have certain knowledge that the Russians only a very few years ago desired to kill more ducks than they had killed in the past. With this end in view they approached certain persons in England to find out methods of decoying and punt-gunning. Such activities, from a commercial point of view, could only have been rendered possible by cold-storage, and the fact that, in Europe, it is legal and customary to import and expose for sale wildfowl taken in foreign countries.

(6) *Silting of Mud in Estuaries, Inlets and old Harbours*

This is a factor which in most cases can only be referred to as one directly due to natural causes such as alteration in the level of the coast-line. But it is mentioned here on account of the part played by the growth of *Spartina townsendi*. The increased growth and spread of this injurious aquatic plant is the result in many cases of the pulling up by swans of their proper food plants, whose place is then taken by the dominating *Spartina*. In a good many inlets along the southern coast of England, swans are present in excessive numbers and the part they play in the spread of *Spartina* is a serious one, as this plant plays an important part in the deposition of ooze and so leads to the obliteration, or alteration in the direction, of the channels and to a widespread formation of underwater banks. Under this section, too, we may refer to the very important part played by *Zostera* as a marine food plant for duck, especially Wigeon, and the necessity for investigating its life-history distribution and diseases (see page 29).

(7) *Punt-gunning and Shooting from Mechanically Propelled Boats*

There can be little doubt that the halcyon days of punt-gunning have gone; and in view of the present developments in the northern breeding areas of Europe there seems every prospect that such a sport will hardly be worth prosecuting in the future, unless conventions of a reciprocal nature are undertaken by the nations of Europe as a whole, combined with a very definite shortening up of the open punting season. At present the occasions when a punt-gunner gets on an average more than one or two good shots in a week or more are few and far between, and he often has to undergo hardships and do plenty of hard work to do this. The present constant disturbance of our estuaries and inlets does not render his task any easier. The total number of duck taken by punt-gunners in Great Britain and Ireland is a steadily diminishing quantity, and this points either to a steadily diminishing supply from the north or to the fact that the duck are more shy and keep to a safer distance out at sea, or to both. Punt-gunning on the continent is generally looked down upon as being unsportsmanlike, and we are considered to be the worst offenders and to be answerable for hetacombs of slain. In all probability this is now quite an unfair view of the situation and an accusation which could not be upheld; but what is almost certain to bring its own unwelcome and untoward result is the continuance of shooting until too late in the spring when Wigeon and other wildfowl are on their way north to breed already paired and full of eggs.

(8) *Disturbance by Aeroplanes*

Whether this disturbance is deliberate and done with the intention of driving wildfowl to shooting parties on shore, or whether it is merely done "for the fun of the thing", it is an intolerable infliction which ought to be stringently forbidden. While a member of a duck-shooting party on some famous reservoirs I have myself seen an aeroplane come down and deliberately and systematically skim over the water with the intention of driving the duck to other quarters, an intention it may be added which was quite successful.

Actual shooting from military aeroplanes at least is now impossible, and it must be extremely rare from civilian aeroplanes.

(9) *The Close Time Factor*

It is clear that if the preservation of wildfowl is to be successfully carried out some decision, based on scientific principles, must be arrived at as to what constitutes a suitable date on which to begin and end the open shooting season. These dates may, of course, vary slightly according to the range and latitude covered by the country in which the dates are determined.

Unfortunately it has for long been the custom in this and other countries to commence the open season too early and go on too long. The result has been, at any rate in this country, that every summer (note the season!) many young wild duck, hardly able to fly, have been shot or even captured by dogs in the reeds; while in the spring, shooting has been continued in spite of the very obvious fact that many birds might have already paired and, indeed, in some cases might have actually begun to nest. Thus at the commencement of every open season the local stock of duck receives year after year a formidable knockdown blow which effectively prevents its attaining the normal limit which local conditions would naturally support. It is also to be noted that this premature massacre of the "young blood" cannot fail in the end to have its repercussions on the virility of the race, for there must be a definite trend for only the old experienced birds to escape while the "new entry" gets wiped out. That this is not a purely fanciful picture is, I think, suggested by the fact that where the commencement of the open season has been postponed until such time as the young birds of the year have a reasonable chance of escape by flight, the stock of wild duck has been doubled or more than doubled. But it is, at any rate, obvious that young pairs will produce larger clutches of eggs than old worn-out pairs.

As regards shooting too late at the end of the open season, the suicidal nature of this policy must be so obvious as hardly to need any verbal protest, but it may perhaps be pointed out that no owner of a properly managed partridge-shoot would be so ill-advised as to continue shooting after it was obvious that the birds had paired, much less had begun to nest. Why then should such a policy be justifiable in the case of wild duck? The extraordinary thing is that, in the case of wildfowl, it is in fact actually done, and, as far as our own country is concerned, with the apparent approval of associations devoted to wildfowl shooting whose members would be both surprised and offended if they were accused of unsportsmanlike behaviour.

Fortunately, however, since the above remarks were written, the passing of the Wild Birds (Duck and Geese) Protection Act, promoted by the British Section of the International Committee for Bird Preservation, has gone a long way to put a better complexion on this state of affairs. But even now the law as it applies to the shooting of wildfowl is not as perfect as it might be, and there is a large body of sportsmen who hold the view, as I do, that it ought to be brought into line with that applying to game so that the open season would begin on September 1st.

II

CONDITIONS IN NORTHERN BREEDING AREAS OF WILDFOWL

The following statements are the personal observations and views of ornithologists and sportsmen who have visited the areas concerned.

North-East Greenland

By C. G. and E. G. BIRD

The area under discussion extends from Kong Oscars Fjord in the south to Hochstetter Foreland in the north between 71° and 76° lat. approximately. A few remarks are added on the breeding of geese further north. Breeding species represented in this area are: Pink-footed Goose, Barnacle Goose, Brent Goose, Long-tailed Duck, Common and King Eider.

Pinkfooted and Barnacle Geese are extremely common breeders; they are found in colonies in all suitable localities. As far as is known this is the only part of the East Greenland coast where these two species are found breeding. Pinkfooted Geese are rarely found further north than 75° or further south than Scoresby Sound. Besides the breeding birds which visit North-East Greenland, great numbers of immature non-breeding birds visit the country every year: this applies to both species.

Factors affecting breeding of geese:

(*a*) *Depredations by Glaucous Gulls, Ermine and Foxes*

In 1938 this was an extremely important factor on Hochstetter Foreland, where there were about 150 pairs of breeding geese. As far as we could see there was not a single goose hatched that year, all the eggs being taken by ermine and foxes. This also applies to the Barnacle but, owing to the fact that it nests on islands and precipitous cliffs, it is probably not affected so much. Eider and Long-tailed Duck also suffer from the depredations of ground vermin.

At the same time it must be borne in mind that 1938 was an exceptional fox year. Foxes, like all Arctic animals, are prone to enormous variations in numbers. In many years they probably have no effect on the geese whatsoever.

(*b*) *Periodic Non-breeding Cycles in the Arctic*

One cannot go into this at any length; suffice to say it has been noted by many authors and is due probably to a climatic or biological cycle or a combination of both.

As far as we know the Brent Goose has only been seen breeding in Scoresby Sound where there is occasionally a small colony. The main breeding grounds are much further north between 77° and 81° N. It is not possible to reach these places with a ship owing to difficult ice conditions, so the birds are only affected by natural factors. Odd geese have been seen by summer sledging parties, who have their base further south. It is not likely in the future that they will ever be affected owing to the almost impassable ice conditions and the fact that there is nothing of economic value there.

(*c*) *The Effect of Man on the Numbers of Geese*

Owing to the extremely difficult ice conditions it is only possible for a ship to reach North-East Greenland in the month of August, i.e. long after the wildfowl breed. There is, therefore, no danger from summer expeditions of any sort. Also, as there is nothing of economic importance in North-East Greenland except the polar fox, it is not likely it will ever be more "opened out" than it is now.

There are about ten stations in North-East Greenland which are manned each year; that means there are about twenty-five people who overwinter. This is obviously very few for such a great length of coast-line. At the time of breeding, which is the beginning of June, there is not much sledging done, as the ice begins to be dangerous at this time. The birds are, therefore, only affected in the immediate vicinity of the stations. Naturally a few birds are shot and eggs taken at the breeding time. The only effect is probably to make many birds move to quieter spots along the coast.

The effect of man on the sea duck we think one can regard as almost negligible. In favourable years great numbers breed. But, as already pointed out, ice conditions make it so that not many eggs can be taken.

In conclusion we think one can say North-East Greenland is still a safe breeding-ground for the three species of geese and sea duck. It is little affected by man and, as far as one can say, it appears these conditions will remain so for many years to come.

Iceland

By MAJOR W. M. CONGREVE, M.B.O.U.

Here, from the beginning of the island's occupation, eggs have been eaten for food. This one can understand and expect. The Icelanders have the sense only to take a proportion of eggs out of each nest found. They know, as do oologists, that in these northern climes the anseres do not repeat a clutch of lost eggs—or at best with occasional exceptions. So far so good, but a thing Mr Freme and I saw was too dreadful for words. The great marsh at Skutustádir, on which vast numbers of duck (and a few Greylags) were breeding, was deliberately flooded (by opening of sluice gates or dykes), in order that the somewhat starved marsh grass should have a good soaking, with a view to making it grow, for hay. Well, they must have hay but it is pretty poor that hundreds of ducks should be wiped out. The damage done by egg-eating Ravens is colossal, as we personally saw. Freme and I have called attention to this in our "Seven weeks in Eastern and Northern Iceland" in the *Ibis*, April 1930. Freme was responsible for shooting over fifty on their morning and evening flights to Myvatn, from their distant mountain-breeding homes. On the evening flight they were gorged with egg, and it was running from their beaks. Now one can say that this has always gone on. I suppose it has, but what has not gone on until, say, fifty years or so ago, is the invention of the breech-loading gun with the everlasting shooting by every Tom, Dick and Harry (or the foreign name equivalents) all over Europe and North Africa. One cannot have it both ways, i.e. destroy the embryonic young and shoot the adults!

Solutions. Try to stop the Icelanders eating eggs by grants of money to get domestic fowls thoroughly established (I may say that fowls' eggs are now being increasingly eaten in Iceland), and try to get the Iceland Government to poison, shoot and trap the Ravens until in more reasonable proportions, and keep them down to reasonable limits. Guns, cartridges and poison should be given out to a number of responsible farmers round Myvatn and other breeding haunts of duck. The geese have, of course, been practically exterminated at Gráfarlönd, where there used to be thousands (see *Ibis* article referred to above .

By S. W. P. FREME

There is no doubt in my mind that geese (Pinkfeet at any rate) must be decreasing, and that while shooting in Britain accounts for quite a large number annually, that number is very small in proportion to the numbers seen, and ridiculously small in proportion to the destruction wrought in northern breeding places. I can produce evidence of this with regard to Iceland.

The colony visited by Major W. M. Congreve and myself in 1929 is actually the first one to be correctly identified in Iceland. We thought there might have been a maximum of fifty pairs. Most nests were at a considerable height from the ground, and it is puzzling to imagine by what means the young birds, from some precariously placed nests, made a safe descent. That large numbers lose their lives in descending I do not believe. Such a mortality is against Nature. That either foxes or falcons, or both, for both were present, were destructive, we found evidence. Indeed no evidence is necessary in the case of the Arctic fox, a sure-footed destructive little brute which must naturally be a menace to ground-breeding birds and particularly to geese. "Particularly" because foxes are now persecuted for the sake of their pelts and because of their depredations among lambs. They have retired before persecution to the wilder inland places and are thus only common in the desolate localities naturally preferred by the Pinkfeet. Even worse than the foxes are some of the farmers. The colony I visited was annually raided by a farmer living about eight hours' ride away from it. How many eggs he took annually I was unable to ascertain. I will give the figures later as to another colony that now seems to be no more! A visit to our colony must have been made in 1930 when I was out in Iceland, for I heard that the same farmer had located another branch of the same colony in a neighbouring gorge. I think he took eggs for food—though he certainly took some that were too highly incubated for anyone but an Eskimo, and possibly some few dealers in egg-shells.

I do not know if his raids continue but have no doubt what they do. As an example of what must happen eventually to all the Pinkfeet in Iceland in time—especially with roadmaking and new facilities for travelling—one may take the deplorable case of that colony which some few years ago was considered strong and flourishing in the Gráfarlönd district. Though the birds here were long supposed to have been *Anser albifrons*, in the light of recent investigations—those of Congreve and myself, my own subsequent ones and the still later ones of Bird and Haigh Thomas—there is little doubt that the birds must have been Pinkfeet. In 1929 I was informed that very few geese were to be seen on the site of the old colony. In 1930, with two guides, one who had visited the district many times in the breeding season in search of strayed sheep, I attempted to go there myself. It worked out at about eighteen hours' ride from where I was staying, and when we were within about six hours of our objective, my guides refused to go further on the grounds that the snow, which was extremely severe on June 10th, might prevent our return.

I had not supposed that I should see many geese even had I got through, but I did not think that the condition of the colony could be as precarious as it will appear from the following translated "questionnaire". This is a record of a conversation conducted (per interpreter) with the householder of Grímstádir farm, about ten hours' ride from the goose colony. I stayed the night to break my journey áfter having turned back from Gráfarlönd. Grímstádir is just over the river Jökulsa-à-Fjóllum, the route taken by me being on the west of that river which I crossed to visit the farm.

This conversation is copied direct from my diary of the "expedition":

Q. Have you been to Gráfarlönd this summer for goose eggs?
A. No! I shall not go.

Q. Did you go last year?
A. No, it would not have been worth while!

Q. Why not?
A. There are so few geese nesting there.

Q. When did you go there last?
A. Three years ago.

Q. Did you take many eggs?
A. Not so many!

Q. When did you last find the trip worth while?
A. A few years before my last one.

Q. How many eggs did you take then?
A. *About six hundred.*

There follow various questions of identification irrelevant for present purposes, all of the answers to which confirmed my belief that the birds seen were Pinkfeet. The last relevant question was:

Q. Why are there so few geese there now?
A. There are foxes in Odádahráun.

Odádahráun is the great lava field, and there are certainly foxes there as I know. But though the farmer was, I believe, really convinced that his own raiding did no harm, I submit that here is sufficient proof of the virtual wiping out of what had been for years a strong colony. Six hundred eggs is a shocking total for a raid and I do not believe that any second clutches would be laid. This raid would have been made about the same time as my intended visit, and there would be little fear of failure as Grímstádir is about eight or ten hours nearer than my base. Supposing the eggs to have been taken on about June 10th, it would have been about July 1st (possibly a week earlier) before the raided birds had got over their recovery period and laid again. A fortnight is a reasonable recovery period to assume, so that the young would not hatch until just about August 1st. At this rate they would not have been able to leave Iceland on the customary date. Pinkfeet used to arrive on the Dee in its palmy days about September 18th.

Gráfarlönd is about 1400 ft. above sea-level, and altogether I do not fancy the chances of a second brood hatching out.

Whether foxes have wiped out the remainder of the colony I do not know. Certainly the majority of breeding birds have left. From my own investigations I suspected that they had gone south to the little-known oasis of Hvánnalíndir on the stream Krégga. I passed this belief on to Brian Roberts who visited that locality after crossing Vátnajokull from the south. He found no geese breeding at Hvánnalíndir, though the country seems to have been suitable, but did find feathers indicating geese had halted there. He could not determine their species. Possibly the old Gráfarlönd colony may now have settled down at Herdubreidalíndir, an oasis not far from the original site. If so, the geese are not much safer than they were before. The point about Pinkfeet is that the species is a very small and local one. Unless there are many colonies yet undiscovered in more northerly regions than Iceland, it rather looks as though the Spitsbergen colonies are about the only hope of increase. The number of suitable oases in Iceland is strictly limited and in time will become less suitable. To find a possible Pinkfoot site in Iceland, one must seek for places inland and away from human habitation, having the following qualifications:

(1) Cliffs of some height—to act as some measure of protection from foxes.

(2) Fresh water—probably the headwaters of some large river or some small stream tributary or distributary.

(3) Grass for food.

Though much of Iceland's interior is little known, there simply cannot be many such ready-made places. Failing some legislation, properly enforced, the Pinkfeet in Iceland stand a poor chance. If the farmers could be made to understand that continual raiding must end in extermination of geese, I believe they might agree to give them their chance. There is a widespread belief that the birds will lay again. Common sense surely argues that they will not.

Whilst persecution in their breeding quarters is undoubtedly—to my thinking—the greatest threat to the majority of species, disease in the same quarters is also a menace. During the summer of 1929 many ducklings of the following species were to be seen dead from apparently natural causes around Myvatnsveit: Mallard, Teal, Tufted and Scaup Duck, Long-tailed Duck, Common Scoter. Probably there were equally extensive casualties in most species.

It may be a coincidence that a disease of extreme severity almost wiped out the birds nesting on Grimsey Island (north of Iceland) in the same year and extended to the rock-breeding birds on the coast near Husavik, yet did not affect the huge Eider colonies at Laxamýri and Thorshovn. It seemed to be confined to Rock Auks and Gulls, though Cormorants were also affected. Probably the diseases were unconnected save by the almost certain common cause of overcrowding. In Myvatn the overcrowding may be regarded as having been aggravated by the flooding I have already condemned. I do not see what can be done about this save that it might be possible for the flooding to be done a little earlier, when at least some nests would be saved, though of course the overcrowding would continue.

I am convinced from my experiences that the only way to safeguard the Pinkfeet is to endeavour to obtain more sympathetic treatment for them in their northern homes. After all one can hardly blame the Icelandic farmer for raiding what he believes to be a common bird. He does not know the harm he is doing, for up till 1929 he did not know that the Pinkfoot bred in Iceland—the very men who ruined the colonies believed that the birds they raided were either Bean or Whitefronts. I think the Icelandic Government would readily co-operate in an attempt to protect this goose, for the Icelander is, I think on the whole, naturally averse to unnecessary destruction of bird life, and most of the offences committed are due to thoughtlessness or lack of accurate knowledge. I do not think any form of subsidy likely to be of use, but I do think that the people are sufficiently law-abiding to respect a Government dictum.

Mr S. W. P. Freme, writing in the *Shooting Times* (April 11th and 18th, 1936), further states as follows:

There are duck in plenty throughout Iceland, and I do not doubt there always will be. Yet to a visitor, whose sole business was the study of breeding wildfowl, one unpleasant fact was all too obvious —the distressing waste of life wrought by the inhabitants, often with the best of intentions, and the destruction permitted either by reason of apathy or lack of observation. In the beautiful district around Lake Myvatn are breeding grounds beyond compare. Here at the height of the breeding season are Mallard, Pintail, Gadwall, Wigeon, Teal, Scaup, Tufted, Barrow's Goldeneye, Long-tailed, Common Scoter, Merganser, Goosander. Or again take the case of that vast expanse of marsh round the snow-powdered hump of Sellandarsfjall. On this wide flat expanse of ground the cover is formed chiefly by growths of dwarf birch and dwarf willow, seldom reaching a height of more than nine inches. Pools abound and the neighbouring villagers can control the amount of water in the marsh by means of culverts. Among the scrub, and in patches around the pools, grows a certain amount of coarse grass, which, in a country where the pony is the chief means of transport, assumes an importance out of all proportion to its quality. To help the growth of this somewhat scanty "feed" it is the custom to flood the marshes for a period, and it is unfortunate for the duck that this period falls at a time when the marshes are teeming with eggs, many of them incubated. Whilst prepared to grant the necessity of flooding, since ponies must be considered, it seemed to me that, had the water-level been raised at an earlier date, many nests eventually flooded would never have been made or laid in. The keynote of the far northern breeding season is haste. Duck and geese, though arriving for the most part mated, have to lay, incubate, fledge the young and moult —the latter resulting in the case of all geese in complete flightlessness—before making the long journey to their winter quarters. A bird drowned out in June has wasted much precious time and has, after the loss of its clutch, to go through an inactive period of some ten days before laying is resumed. The flooding does not, of course, submerge all nests, though it does destroy hundreds and possibly thousands of eggs. All nests on higher ridges of ground escape, and, after the recovery period, these are augmented in number until every small flood-formed island is packed with nests. These, quite unreasonably, are raided by villagers for eggs to be eaten immediately or preserved for the winter. This raiding, at this time, is an abuse. Since so many nests were to be flooded, one would have supposed that it would have been reasonable to raid those that were on low ground instead of those that had survived. In any case most of the eggs taken in 1929 were so highly incubated as to be unfit for consumption by even the most broad-minded. This practice of flooding, among other evils, locates the surviving nests for the countless marauding Ravens. It is not my purpose here to launch an attack on the Iceland Raven, but I had considerable experience of the damage done by this species and shall be delighted to go further into the matter with anyone who is so shortsighted as to compare our own rather scarce bird with the very common and entirely uncontrolled one in Iceland. I will confine myself to the statement that although the "locals" in Myvatnsveit were very pleased with the merciless campaign against Ravens that I conducted in 1929 and 1930, they made no effort to check or control the birds themselves. The Raven, though often a menace inland is, on the sea coast, not only harmless but useful. There is much garbage to be disposed of at fishing ports.

Spitsbergen

By MAJOR W. M. CONGREVE, M.B.O.U.

It was in 1932 that I was in Spitsbergen. To my personal knowledge a colony of Pinkfooted Geese was raided by egg-eaters and, when I went there, only two nests had survived. One because it was ungetatable without a rope and the other only held one or two eggs. We were told that over 100 eggs had been eaten. Various other spots visited had been well cleaned up. The one and only easily reached colony of Barnacle Geese near Longyearbyen, on Ice Fjord, had been raided as regards the easily reached nests—for food. The colony has since been shot out. No particular matter, as regards this species, as probably it breeds in peace in Greenland, Novaya Zemlia and elsewhere in Spitsbergen, etc.

However, the Pinkfooted and, I may say, the Brent Goose are another matter. These birds, as regards Spitsbergen, nest in easily reached islands, floors of valleys and easy terraces or bluffs. They are now cleaned up unmercifully by sealers who, I am told, export many of the eggs to Norwegian markets at Hammerfest, Tromsö, etc. Steward tells me that on islands where Brent bred freely a few years ago, none are now to be seen. He met one woman on an island, who said that she had taken it for the eggs and resented anyone landing to look for them.

Solution proposed. Norway cannot prevent inhabitants and sealer crews from eating fresh eggs in this out-of-the-world spot. They can prohibit the sale of eggs in Norway. It is to be hoped that some geese breed on the ungetatable east side of Spitsbergen—there they are at present safe, as no boat can get there in the breeding season and I doubt serious attempts to colonise the east side of the island. The Norwegian Government should also, surely, be able to prevent the renting of islands for their eggs.

By C. T. DALGETY

My experience of Spitsbergen consists of expeditions in 1927 to Edge Island, and in 1930 and 1931 to the west and north of Spitsbergen and to North-East Land.

1927 appeared to be a bad breeding year without human intervention.

The Eiders on the small holms to the south-west of Edge Island had had nearly all their eggs and down taken. The few ducklings seen were very young in the beginning of August, showing that no first broods had been allowed to hatch. Only two broods of Pink-footed Geese were seen amongst about thirty old birds. No young Brent Geese were seen, though over a thousand old Brent were seen. Eider hunters were still in Stor Fjord as late as the last day of July. A sealing boat, spoken to on the last day of July, had on board, besides 1000 seals, fifteen bales of Eider-down and many barrels full of eggs. Hanging from the boom were a number of Brent Geese. The crew of this boat also killed at least six reindeer, whose freshly cut-off heads were seen by us lying on the ground. The Spitsbergen reindeer is protected by Norwegian law, but I have no hesitation in saying that the crews of all sealing boats take every opportunity of killing all the reindeer that they see, excepting only in Sassen Valley which they consider to be too near Longyearbyen, and consequently the risk of their being found out too great. The trappers who winter in Spitsbergen do not seem to kill the reindeer in the summer, but I think that they do so in the autumn. However, these men do need them. The reindeer of Spitsbergen is likely to be exterminated unless the Norwegians enforce their laws. As this reindeer is a distinct breed peculiar to Spitsbergen, it would be the greatest pity if it were exterminated.

1930 and 1931 were both good breeding years for most birds. Eider eggs were being collected up to very much later than June 10th, which I believe is the date after which it is against Norwegian law to take Eider eggs. The Eider hunters, or at least many of them, take all the down from the nests, leaving the eggs to be incubated on the frozen ground. At least that is what they used to do in 1930, but I am told that now they take all eggs, even those which are very much incubated. Of this I have no personal knowledge.

The Norwegian is sometimes criticised by other nations as being a destroyer of wild life, but he only kills those things which he can make use of. The sealers, as a class, are very poor and have a hard time every winter, therefore they like to lay in as large a supply of food as possible. What is needed is to convince them that they must not decrease their breeding stocks by shooting in the breeding season and by taking all eggs that they can lay their hands on. The present law prohibiting the taking of eggs after June 10th cannot possibly be enforced in Spitsbergen, but a law prohibiting the import of eggs into Norway after a certain date could be enforced.

By CAPT. J. H. McNEILE

The position of the Brent Goose, *Branta bernicla hrota* (Müller), in Spitsbergen is becoming increasingly precarious, and the only way it can be made secure is by giving a reasonable measure of protection also to the Northern Eider, *Somateria mollissima borealis* (Brehm).

For the danger from foxes has driven the great majority of Spitsbergen Brent Geese to nest in company with the Eiders on the various suitable islands round the coast, rather than on the "mainland". Unless the ice conditions are unusually severe, the two species find a safe refuge from foxes on these islands, but the very fact that they have been driven to concentrate together in such colonies makes them all the more liable to molestation by the mining population and trappers of Ice Fjord and by the crews of passing sealing-sloops, and *especially* to the far more destructive raids of the "hunters" who come up for the express purpose (we were told by our own Norwegian skipper) of gathering eggs and down for sale in Longyearbyen and in Norway.

There may be anything from one or two to as many as perhaps fifteen Brents' nests among hundreds of Eiders' nests on an island, but no distinction is made by any of these marauders between the nests of the two species, although their eggs and down are both, of course, easily distinguishable.

The Brent Goose is particularly vulnerable to persecution, however, for not only does it lose its normal fear of man during the breeding season, and thus fall an easy victim to any unscrupulous gunner, but *if its eggs are taken, it appears only very exceptionally, if indeed ever, to attempt a second clutch*—unlike the Eider, which will lay again and again, until it probably eventually succeeds in rearing a brood.

It would be interesting to know whether the 1800 Russian miners in Green Harbour pay more attention to the Norwegian Bird Protection Laws than do the Norwegian mining colony at Longyearbyen. But these men cannot reasonably be blamed if they take eggs for food from the easily accessible Eider colonies in Ice Fjord, such as the one on the Anser Islands, for instance, for there are no chickens in Spitsbergen yet, and Eiders' eggs make excellent eating. Moreover, we found that they generally had the good sense to take only *fresh* eggs, which would do little harm if they restricted themselves to Eiders' eggs. The numbers of these duck breeding on the Anser Islands in 1938 appeared if anything to be larger than we saw there in 1934 or 1935.

But unfortunately the Brents' eggs get taken as well, with the result that wherever we went in Ice Fjord in 1934, 1935 or 1938, the only Brent Geese we saw were flying about in parties—birds which had lost their eggs and which therefore had no hope of rearing any young that season. We several times counted as many as thirty birds in such a party.

The "hunters" who come from Norway are a very much worse menace, however, for although they do not appear to harry the islands in Ice Fjord (possibly on account of the miners, or because the Norwegian Governor of Spitsbergen has his headquarters—and a powerful motor-boat also!—at Longyearbyen), the wholesale and indiscriminate destruction they work elsewhere is appalling, and as inexcusable as it is senseless and shortsighted. The difficulty is to know how to check their irresponsible and devastating activities before it is too late, on a coast-line like that of Spitsbergen, where it is virtually impossible at present for the Norwegian authorities to enforce their Bird Protection Laws.

There are not many of these "hunters", but the damage they do is out of all proportion to their numbers. For they make their way from one suitable island to the next, systematically clearing each in turn of all eggs, without discriminating between Brents' and Eiders' eggs, nor even between those which are fresh and those which are incubated.

On July 2nd, 1935, to quote only one or two examples of what is going on, we visited the Station Islands in Liefde Bay, and found a perfect swarm of Eiders nesting on them; in fact as we rowed back to our sloop the largest island looked literally *white* with the drakes keeping their sitting ducks company, the clutches being in all stages of incubation.

The following morning we saw a small Tromsö sloop anchored off these islands, and when we returned to them that evening we found they had been cleared, every single egg had gone, even Grey Phalaropes' eggs had been taken!

Two days later we landed on the Mouettes Islands a few miles further south, to see whether the Tromsö men had been there too. They *had.* And they had evidently taken twenty nests of Brent Goose there, for forty of those birds rose together from a marshy hollow in the middle of the larger island, and we came across three of their emptied nests with scraps of down still in them, although we searched carefully over the whole island without seeing anything of either a sitting goose or of any anxious gander in attendance. It was only too obvious that not a single nest of theirs had escaped the raiders.

These islands are the nesting-place of innumerable Eiders too (or they *used* to be in 1930 and 1931), but we failed to find more than an odd nest or two of theirs, all the rest had been gathered wholesale. Every one of the Brents' eggs must have been hard-set, and very few of the Eiders' can have been fresh so late in the season as that. And no doubt this same sloop went on to harry the Lerner and the Canards Islands in the same way, to mention only those in Liefde Bay. But what did these men do with all the hundreds of incubated eggs they had taken? And how long can the Brent Geese survive such treatment? It seemed more than doubtful whether one single gosling could have been reared that year on any of the Liefde Bay Islands, the principal stronghold of the species in northern Spitsbergen, as a result of the ignorant ruthlessness of this one Tromsö crew.

The Norwegian Government may perhaps have realised the danger, for we found in 1935 that they had started leasing some of the islands on the west and north-west coasts to people who had come up from Norway for the purpose of farming the Eiders' eggs and down during the nesting season.

In 1938 we saw several new huts established along the coast, and their occupants appeared to be farming the Eiders on adjacent islands systematically and reasonably. The only ones we spoke to were a party of trappers living near Cape Roos just south of the Mouettes Islands, but they did not tell us whether they held any lease or special privileges from the Government.

If a system of leasing islands or of scientific Eider farming could be extended, and if these Eider farmers could not only be taught how to distinguish the eggs and down of the Brent Goose from those of the Eider, but induced by some means to spare the Brents' nests and to refrain from shooting the birds at their nests or while they are moulting and unable to fly, the situation could be saved.

They would collect enormous numbers of fresh Eiders' eggs for sale in Longyearbyen or in Norway, as well as great quantities of their down, but they would take good care in their own interests (as the Icelanders do in the same circumstances) to allow the ducks to hatch out a proportion of their eggs, so as to maintain their breeding-stock. And they would presumably see to it that there was no poaching on their preserves by unauthorised marauders.

If, however, it is asking too much to suggest that they should leave their geese unmolested, the only alternative remedy would seem to be the establishment of a sanctuary somewhere in Spitsbergen.

Liefde Bay, with its groups of islands, would be ideal for the purpose, but in years when the ice allowed access to it, trustworthy watchers would have to be established as early as possible during the nesting season either on Reindeer Peninsula or on Grey Hook, from which they could observe any vessel entering the Bay. Norwegian sealing-sloops and fishing boats carry a conspicuous identification number.

In some years such as 1935 the severe ice conditions prevent any approach to Liefde Bay before the second or third week in July. But if the ice cuts the islands off from human raiders, it is all too likely to enable foxes to reach them, and one fox on an island, as we saw for ourselves that year on Hermansen Island and again on Gerd Island, will prevent a single Eider or Brent from laying.

Natural conditions vary so much from season to season in the Arctic that it is often difficult to estimate or compare the numbers of such birds as Brents or Eiders.

But whether or not the latter are managing to maintain their numbers, our experiences in 1930, 1931, 1932, 1934, 1935 and 1938 showed a definite and marked reduction during that period in the numbers of the geese round the western and northern shores of Spitsbergen, and the Brent Goose seems indisputably and inevitably doomed to become a rare bird up there unless steps are taken without delay to protect its eggs and to prevent the bird itself from being slaughtered during the breeding season or moult.

By DR E. S. STEWARD

I made expeditions there in 1931, 1932, 1934, and 1935. In 1931 there was only one settlement, viz. Longyearbyen in the Ice Fjord, inhabited by Norwegian miners and officials. They collected thousands of Brunich's Guillemots' eggs for food and the range of cliffs where they nested was worked by an Estonian. Now, that is, since 1932, this Estonian is engaged at the Russian mine in Green Harbour where there are Russian miners. But the Anser Islands, two or three hours by motor-boat from Longyearbyen, which in 1931 swarmed with Eiders and a few Brent Geese, are now sadly harried. In 1935 I saw a few birds, but not an egg. The inhabitants of Longyearbyen take the eggs for food and the down goes to Norway. There is a Norwegian living on Danes Island at the extreme north-west of Spitsbergen. He rents two or three islands from the Government and farms the thousands of Eiders that nest on his islands, but he always leaves an egg or two in the nest for the duck to hatch.

In Liefde Bay are many islands inhabited by Eiders in vast numbers, Brent and Pinkfooted Geese.

In 1934 my boat was the first there, and it was an amazing sight to see all the birds, but after two or three days I heard a motor-sloop, and my skipper told me "it was the hunters". The eggs of every Eider and goose, together with the down, disappeared. This sloop makes the journey regularly, for I saw it also in 1932 when it was there before me and all these wonderful islands were barren. The birds were there, but all nests and eggs were gone. I am told thousands of ducks' eggs go back to Norway for food, but how many pounds of down is collected I do not know.

1935 was a bad year owing to the ice which stopped me at the North-West Point off Spitsbergen. Several islands on the west coast which were favourite nesting places for Eiders and Brents were deserted, the reason being that foxes had become marooned, and these animals ate any eggs they could find and naturally frightened the birds away. Other islands had been visited by Norwegians, because we found nests empty. Parties of bereaved Brents were flying restlessly about, and we found heads and feathers of these geese which had been shot a few days previously.

It would be extremely difficult to prevent this wholesale collection of eggs, for there are so many remote and distant islands or flat areas where the birds are.

I think that there is no doubt that in Spitsbergen the Brent Goose is being steadily reduced in numbers.

Norwegian Lapland

By MAJOR W. M. CONGREVE, M.B.O.U.

My experience in Finmark (two seasons) is that man does little harm. I understand that Norwegians do not now shoot (I saw no man with guns or signs of shooting) the Lesser Whitefronted and Bean Geese from their nests—according to Steward they certainly used to do so. However, the nests are scattered over vast areas and are in most cases extremely hard to find. Inaccessibility of the breeding home will always prevent man doing much harm. True, roads are now being made. I saw no eggs for food in hotels, etc., as in Iceland. Hen's eggs were supplied everywhere I went. The eggs of the Golden-eye and Smew (mainly the former) are sometimes collected for food from artificial nesting-box sites, but since neither species has real meat value they are rarely killed and their status does not appear to be endangered.

III

THE DISTRIBUTION OF *ZOSTERA* (EEL GRASS, WIGEON GRASS) AND OTHER SEASHORE PLANTS IN RELATION TO THE MIGRATIONS OF WILDFOWL

By Dr R. W. BUTCHER

In 1932 attention was called to the common and well-distributed marine flowering plant known variously as grass wrack, eel grass, wigeon grass, because it had decreased with comparative rapidity along the Atlantic coasts of America, in the north of France, in Denmark and in many parts of England. The reason for this sudden decrease is as yet not known with certainty, and each writer on the subject appears to stress a different cause. These will be discussed later.

One of the most striking facts about the dying away of *Zostera* was that (in England) the largest plants were most affected and later, where the beds recovered, it was a smaller-leaved plant that became most abundant. There are, in fact, three different forms of *Zostera* growing on the seashore and, though intermediate plants are found, the three main varieties are fairly easily distinguished (see Butcher (4)).

The largest form is referred to as *Z. marina* L. and has leaves 4–5 mm. wide and 30–60 cm. long with three to seven veins in the leaf.

In the second form the leaves are much narrower (1–3 mm. wide), smaller and with one to three veins, and such forms are referred to as *Z. marina* var. *angustifolia* Hornem. Recently Tutin (10) has suggested that this form should be given specific rank and has named it *Z. hornemanniana*. In addition to the narrower leaves, he gives as differences a stigma only half as long as, instead of as long as, the style, and the seed 2·5 mm. long, as against 3·5 mm. long in *Z. marina*. These appear to the author to be extremely weak taxonomic differences, and together with the facts that forms intermediate between *Z. marina* and its var. *angustifolia* are often very abundant and the chromosomes of the two forms are identical (Blackburn (2)), it would seem the best to retain for these smaller-leaved plants the accepted status of a variety.

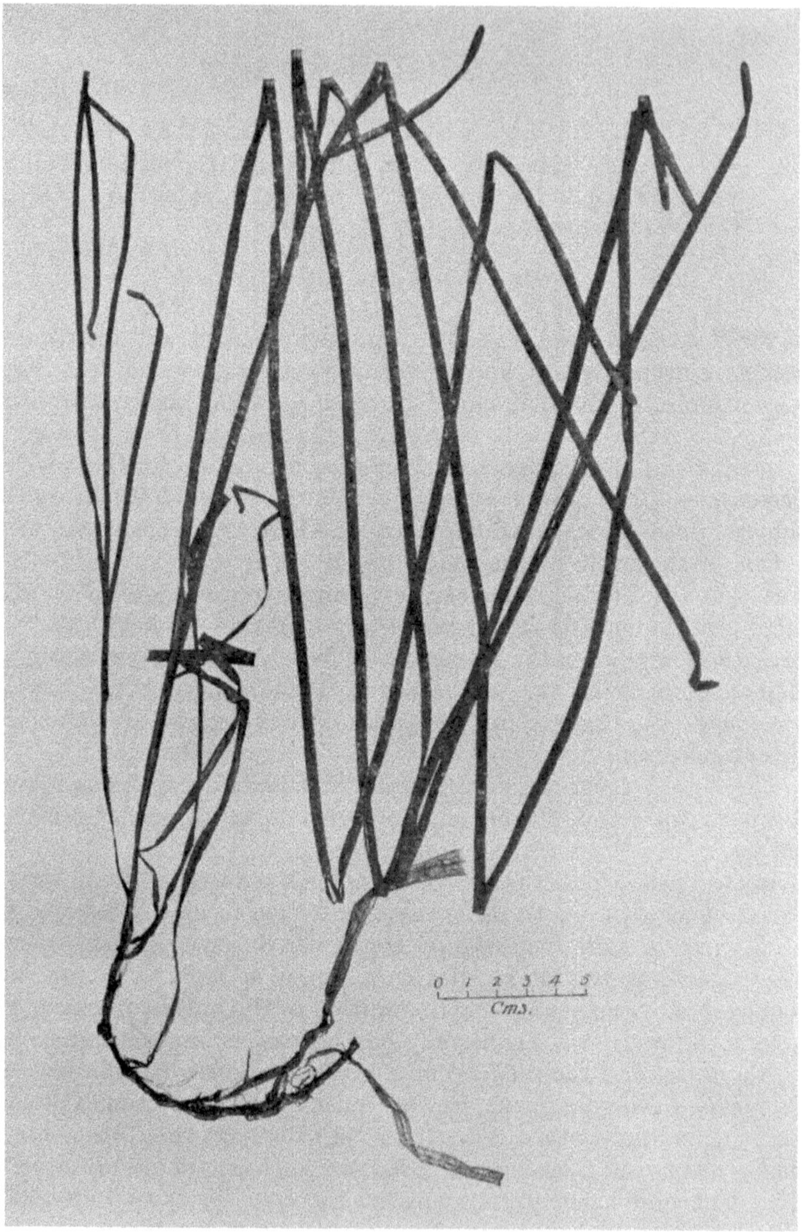

Fig. 1. *Zostera marina* L.

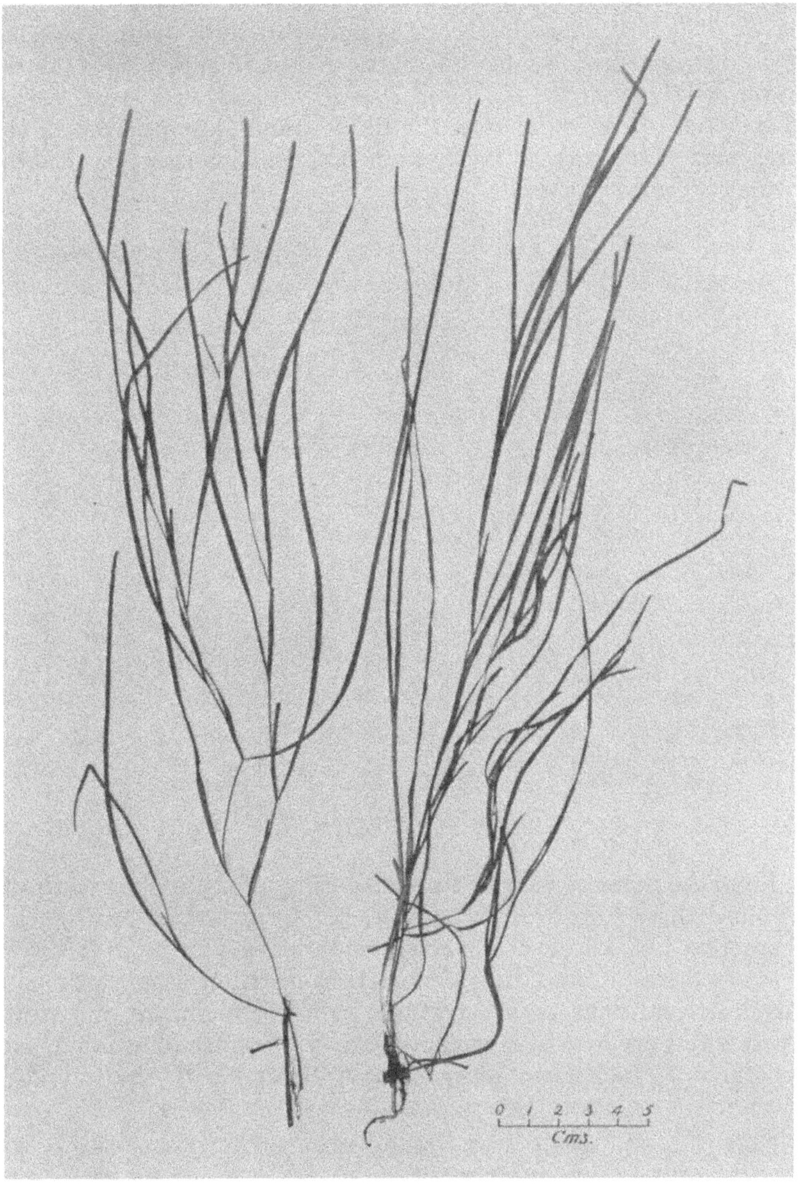

Fig. 2. *Zostera marina* var. *angustifolia*.

A third form is usually smaller and has very small one-veined leaves. This is *Z. nana* Roth and is undoubtedly a distinct species. It is distinguished by having extra-marginal appendages to the flower spike (these are absent in *Z. marina*) and, in a cross-section of the stem, it is found that the fibre strands are situated in the inner part of the cortical layer, while in *Z. marina* they are situated in the outermost cortex.

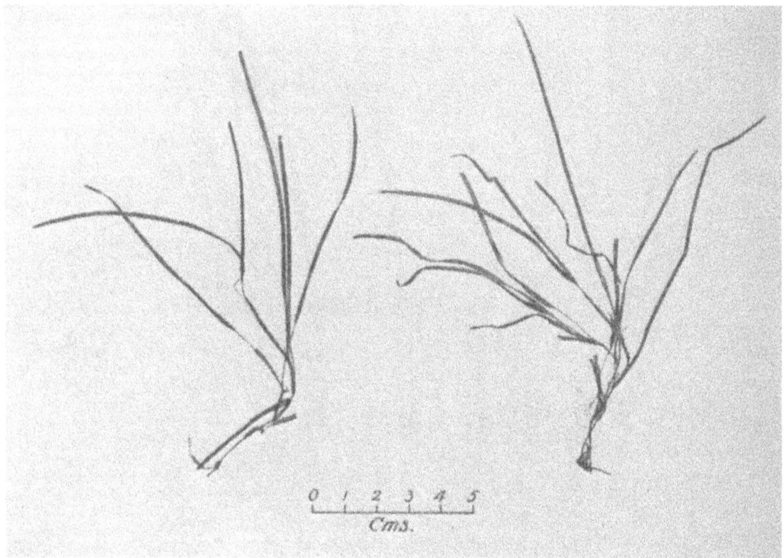

Fig. 3. *Zostera nana* Roth.

From the point of view of the relationship of these plants and the wildfowl these forms are important, not by reason of any specific characters but because of their comparative size, and therefore their food value, and because of their distribution according to physical conditions of substratum, light, temperature and tides. These forms are here distinguished simply by width of leaf and, as a general rule, such forms may be co-ordinated with the following species:

Leaf 10–4 mm. wide	*Z. marina* L.
,, 4–1 mm. wide	*Z. marina* var. *angustifolia* Horn. = *Z. hornemanniana* Tutin
,, 1 mm. wide	*Z. nana* Roth (other characters distinguish this species)

Distribution of Zostera

It is somewhat difficult to obtain a "time picture" of the distribution of the three forms of *Zostera*, especially for the past, as records and memory are unreliable and very few observers have distinguished between the three forms. The map on p. 34 indicates what may be termed the "maximum distribution", that is, all the areas are shaded where any form of the plant has been found at any time. Areas are also marked where it is certain the plant has largely disappeared. In Ireland the plant is recorded for every part of the coast.

The map shows that the grass wrack is by no means universally distributed and that there are large areas where the plant has not been recorded. These are chiefly the wave-swept, shingly and rocky shores on the west. On the other hand, it is, or was, very abundant in the large, sandy, wind- and storm-protected areas of the eastern side and also in the long lochs and protected bays of Ireland and western Scotland.

As regards the three forms *Z. marina*, *Z. angustifolia* and *Z. nana*, there is found to be a difference in distribution according, chiefly, to the nature of the substratum and the time of exposure to the air after the tide recedes. Presumably also the temperature and the intensity of light have some effect, though, with turbidity and height of the tide constantly varying, these last factors are very difficult to assess.

The principal cause determining the distribution is exposure to the air and, therefore, to desiccation. Plants growing near low-water mark are exposed for only a very short time, while those near high-water mark are exposed for nearly as long as they are submerged. In this connection the water-retaining power of the substratum is also important, and whereas sand loses water comparatively quickly, soft mud loses it more slowly, and plants growing in the latter places, especially if in watery depressions, can stand considerable periods of exposure. The large-leaved *Z. marina* grows chiefly in the places where there is little exposure and on sand rather than mud. It is also found in sheltered salt lagoons where there is no variation of water-level. The small-leaved *Z. marina* var. *angustifolia* is commonest in muddy estuaries up to about mid-tide mark. It also may grow in lagoons, but always in mud and not sand. As well as exposure, the movement of the substratum is important.

Fig. 4. Map showing distribution of *Zostera*. Places where it has become extinct are marked with a circle. The data for Scotland are incomplete and are based on topographical botany. The data for England are based on county floras and personal observation.

Where a sand or mud bank is unstable, it is uninhabited and it is stated (12) that there is 68 per cent of such sand in Liverpool Bay. In places like the Wash, therefore, *Zostera* is absent because the sandy substratum is too unstable for it. There is also to be considered the chemical nature of the substratum, as it has been shown that flowering plants absorb food substances through their roots and not from the surrounding water (Pond (8)), and there is no reason why *Zostera* should be an exception. Tutin's (11) statement to the contrary is backed by no sound experimental evidence. It is thus mainly exposure that limits the area of growth and the size of the plant: in the little exposed places at low-water mark and in lagoons, *Z. marina* is most abundant: with more exposure and mud rather than sand *Z. marina* var. *angustifolia* is dominant, and the upper limit of growth is determined by excessive drying out and low salinity. *Z. nana* is rarer than *Z. marina* var. *angustifolia*, but occurs in similar places.

The History of the Wasting Disease of Eel Grass

As mentioned previously, between 1931 and 1933 reports came from various parts of Europe and the Atlantic coast of America that eel grass was disappearing rapidly, and wildfowl were migrating to different feeding grounds.

In England the first reliable reports came from the Blackwater Estuary, Essex and the Fleet in Dorset. As a result of an inquiry by the author in 1933 (Butcher (3)), the following important facts of the disappearance were established.

(1) So far as England was concerned the disappearance of *Z. marina* had been noted long before 1932, and the consensus of opinion was that dying out was first noticeable in the period 1920–2. There were, however, the reliable reports mentioned above which show there must have been an obvious decrease larger than usual in 1931 and 1932, but it is erroneous to suppose that the decrease first occurred in 1930 and 1931 (see (11)).

(2) The decrease was noted to be chiefly with the large-leaved *Z. marina*, though exceptions are recorded (Lynn (5)). Whether the narrower-leaved forms were affected to a smaller extent or not at all still remains in doubt. The important point is that, at the moment, *Z. marina* var. *angustifolia* is the common plant and *Z. marina* the rare one.

(3) Decrease was not at all regular. Some places were affected badly, some not at all. The following is a brief summary of the conditions of *Zostera* based on the author's survey of 1933 and other data kindly supplied by various observers.

Northumberland. Quantities of *Zostera* occur in four places and most abundantly at Fenham Flats, and there has been no change or a slight increase in the total volume, except one year when the beds were extensively damaged by the oil from a burning vessel. The plant is almost entirely *Z. marina* var. *angustifolia.*

Durham. There are records only for the mouth of the Tees, and here it is now almost extinct. The Tees estuary has been extensively dredged, and large areas of mud flats have been reclaimed for industrial purposes.

Yorkshire. Z. marina var. *angustifolia* is found on sandy mud on the northern side of the Humber, and a full account of this is given by Philip(7). He is of the opinion that substratum largely determines its distribution.

Lincolnshire. Zostera is very rare on the Lincolnshire coast presumably because of the lack of satisfactory stretches of shore of the right type. It is especially worthy of note that there is none of this plant in the Wash and its absence is attributed to the unstable nature of the sand.

Norfolk. Z. marina, Z. marina var. *angustifolia* and *Z. nana* are all found in the sheltered inlets of the north coast, but not very abundantly. There is circumstantial evidence of a gradual diminution for many years at Wells and in Breydon Water.

Suffolk and Essex. Southwards to the Thames mouth *Zostera* becomes increasingly abundant in the sheltered bays and estuaries of Suffolk and Essex. All the three forms are found growing together in the Blackwater, where there was a noticeably sudden decrease in 1932, and in the Stour. There is no doubt that the plant has almost disappeared in the Orwell and that this decrease has been going on for many years.

There is a salt-water lagoon at Walton-on-the-Naze which contains a pure growth of *Z. marina.* Up to 1932 it was the practice of the owners to let out the water in the winter and kill the plant by desiccation. This kept down the quantity. In 1932 so much had died that this was not necessary. It should be noted, however, that no other form has replaced the large form which, after four years' scarcity is, in 1939, as abundant as ever.

Kent. *Zostera* is now almost extinct in the Thames and Stour estuaries, but its diminution would appear to have been going on for at least twenty years and is due more probably to the alteration of currents and sandbanks than anything else.

Sussex. There is a little *Zostera* to be found in the estuaries Cuckmere and Arun, and it is more plentiful in Pagham and Chichester harbours. The narrow-leaved forms predominate and *Z. marina* is seen but rarely below low-water mark.

Hampshire. *Zostera* used to be abundant in all the estuaries of the county. It is now almost extinct in Southampton Water owing to, among other things, the building of new docks and the dredging of the channels. It is almost extinct in the Lymington river. On the other hand, all forms are abundant at Ryde and other places on the northern shores of the Isle of Wight. There is evidence of a diminution of the large form between 1925 and 1927. All the estuaries of Sussex, Hampshire and Dorset have been considerably altered in recent years by the advent of *Spartina townsendii* (see p. 46).

Dorset. The only two sheltered places in this county are Poole Harbour and the Fleet behind Chesil Bank. There has been a gradual decrease in Poole Harbour, while in the Fleet there was a very heavy and obvious mortality of the large *Z. marina* in 1932. In 1933 and subsequent years there has been an increasing volume of *Z. marina* var. *angustifolia* rather than *Z. marina*.

Devon and Cornwall. All three forms have been found in some abundance in the protected harbours and estuaries such as Salcombe, the Yealm, Marazion, etc. Where it was observed, the history of the decrease seems to have been very similar to that in the Fleet: sudden decrease in 1932 and a gradual recovery in the quantity of *Z. marina* var. *angustifolia*. There is no record of the plant in North Devon; in North Somerset it is very rare and it is absent in the Severn estuary.

Glamorgan. It is of particular interest to note that there has been an increase of *Z. marina* var. *angustifolia* in recent years on the coast of Glamorgan, and the inshore fishermen complain that its increase will ruin the shell fisheries. In the Blackwater, Essex, the fishermen complained that the disappearance of the "long-weed" would ruin theirs.

Wales. For the rest of the western coast there are records of *Zostera* only in the Menai Straits, the river Dee and the Solway Firth.

Scotland. The data for Scotland are very sparse and cannot be laid out in such detail as for England. In particular there are no records of sudden disappearance in 1931 or 1932, nor can one say for certain what forms occur. *Z. nana* is certainly very rare, but plants with leaves of all widths have been collected on the western coasts. Records of decrease come from the North Uist (Hebrides) and, on the eastern side, *Zostera* has been decreasing gradually for many years in the Tay. It is still abundant, however, in Cromarty Firth, but what varieties it has not yet been ascertained.

Ireland. Zostera is generally distributed, but there are not sufficient data to map out the varieties. Diminution was recorded from Cork in 1931 and 1932 and from Strangford Lough in 1926. The history at the latter place has been well worked out by Lynn (5), and the following points are worth summarising:

(1) The plant most affected was a plant with leaves 3 mm. wide, i.e. not the largest nor yet the smallest.

(2) Search for the two parasites recorded elsewhere (p. 39) as causing the disease failed to identify them.

The Cause of the Wasting Disease

In spite of all that has been written on the subject, or perhaps because of the many conflicting facts and statements, the cause of the disappearance of *Zostera* still remains to be found and there seems to be no known method of controlling its quantity. It should be borne in mind that in searching for the reason we are dealing with a series of contributory causes rather than a single catastrophic event.

Briefly, the causes that have been mooted are: (1) winter killing by ice; (2) too high a summer temperature; (3) increase in salinity; (4) destruction by waves and storms; (5) consumption by geese and swans; (6) destruction by competing plants; (7) excessive commercial harvesting; (8) destruction by oil or other pollution; (9) disease caused by a fungus or other living organism; (10) change in the physicochemical factors involved in the relationship between the plant and its surroundings.

It would seem that the disappearance in different places is due to different causes or to a different combination of these causes. Causes 1–4 may be ruled out; rarely does the sea freeze, the summer temperatures were not particularly high; the plant can grow in water 50 per cent more saline than the sea and storms would not affect a large number of places, e.g. the Fleet, where the plant suddenly disappeared.

Many observers have recorded that Mute Swans tear up the plant by its roots and are very destructive on the beds and it is conceivable that locally they may help with the decrease. They are not, however, a primary cause, as swans and geese are known not to have fed extensively on some places where the plant has disappeared. With the present shortage it is important that where the Wigeon and Brent have concentrated on remaining *Zostera* beds the quantity should be carefully watched and recorded.

As regards destruction by competing plants, this may have been the cause in a few limited areas and the chief competing plant is *Spartina townsendii*. The inter-relationships of this plant and the physical conditions in which it grows are dealt with on p. 46.

Excessive commercial harvesting may be ruled out, as also can oil except on very rare occasions.

Disease is considered by many workers as the prime reason for the diminution, though they are not agreed as to what causes the disease. It should be emphasised first that there can be two types of disease; an infectious disease caused by a living organism and a diseased state brought about by some change in the physiological state of the plant. Disease, therefore, covers nearly everything; it is the cause that is difficult to find.

A bacterial disease seems now to be ruled out by recent work. Two fungus diseases have been recorded, *Ophiobolus* by Mounce & Diehl (6) in Canada and a *Labrynthula* in the United States by Renn (9). Both have been recorded by Tutin (11) from this country, the latter on *Z. marina* only, and not on the other two varieties. This author does not accept these parasites as the sole cause of the sudden and widespread diminution of *Zostera*, because there is no evidence that the infection spread rapidly from one region to cover the large area now affected, and it is unlikely that two independent organisms became suddenly virulent at the same time. His contention is that these parasites have been living in *Zostera* for some time, and a sunshine deficiency in 1931–2 checked the vigour of the plants and as a result the parasite got hold and did a lot of damage. Atkins (1), however, showed that the sunshine deficiency in 1931 was not abnormal and that even then the amount of sunshine at Plymouth was equal to that of a normal year in the north of Scotland where *Zostera* still flourishes.

Another matter not made clear is the relationship of *Z. marina* and *Z. marina* var. *angustifolia*. The former is now rare; it is doubtful whether it was ever very common. The latter is the common form in all Britain, i.e. if all intermediate forms are included. The parasites attack *Z. marina* chiefly and only rarely *Z. marina* var. *angustifolia*. Yet there has been, in many places, as much diminution in the latter as in the former and Lynn (5) states that in Strangford Lough the latter is the form mainly affected. All these contradictions certainly make one feel that a parasitic disease is not the only cause, or indeed the main cause, of the diminution of *Zostera*.

Finally, then, there is the suggestion that a change in the nature of the substratum is one of the principal causes. The ways it could change are many, and there is plenty of evidence that in some places that change has taken place: in others there is no evidence. It is certain that *Zostera*, along with all other organisms, will not live in shifting sand, and there is a very large proportion of such uninhabited sand in all estuaries. In the Mersey (12) it is 68 per cent. Further, by deepening channels, by building training walls in an estuary, the currents may be changed and firm sand may become shifting. Such a cause should be looked for in highly commercialised and dredged estuaries such as the Thames, Southampton Water, the Mersey, the Clyde, and possibly the Tay and Plymouth Sound. If the burrowing animals have also disappeared from where the *Zostera* beds were, then that is very good evidence that the substratum has become impermanent.

Then there may in some places have been changes from sand to mud. It is accepted that *Z. marina* grows in sand and *Z. marina* var. *angustifolia* in mud. A sand to mud change might well cause the diminution of the large form and increase of the small forms. Cultivation experiments to determine whether size of plant is in any way related to the nature and composition of the substratum are sadly needed. A distribution of *Zostera* according to the substratum is also given by Philip (7) and is supported by the observations of Lynn (5), Blackburn (2) and others.

With all this evidence it would seem that the nature of the

substratum may be an important, though not the only, cause of the diminution of eel grass, but evidence is very sparse and the wasting disease is as much a mystery as ever. It is also quite possible that, as Tutin(11) suggests, the parasite is always there, and a change in some factor which causes a weakening of the plant will allow the parasite to become virulent and wipe out large areas. Though lack of sunshine has been shown not to be such a factor, others, including the substratum, may be. It is rather curious that Tutin, in his paper, ignores the substratum almost entirely.

In concluding, it would seem that a disease and the change in the substratum are two very likely causes, and the following questions, if answered, would help to solve certain aspects of the problem.

(1) What is the present distribution of small and large forms?

(2) Does the size of the plant vary from season to season and can cultivation increase the size?

(3) Where *Zostera* is dying, is there a change in the numbers of burrowing animals showing a change from permanent to impermanent sand?

(4) Which plants produce fertile seed and in what circumstances?

Relationship of the *Zostera* Beds to the Numbers of Migrating Wildfowl

It has been stated by many competent observers that wildfowl and, in particular, Brent, Wigeon and Swans, feed largely on *Zostera* and their numbers are dependent on the quantity of feed. This is a broad generalisation that is only partly true. If it be correct that these birds are wholly dependent on *Zostera* for their food, they will have before now left the places where the plant has died out, they will exterminate the small quantity remaining in the denuded areas and then concentrate on the few remaining places where there is still plenty and reduce seriously the quantity there. Then they will die of starvation or disappear from the coasts. Such an outlook, if it is the correct one, is very black indeed, but fortunately there is no reason to suppose that it is invariably the case that these birds are dependent on *Zostera* alone.

They are known to be abundant and to feed in places where this plant does not occur. This is confirmed by Dr J. W. Campbell from a preliminary examination of the stomach contents, and much more information on this aspect would be very valuable.

A map is presented of the distribution of eel grass, and it is most important that this should be brought up to date and the omissions filled in, more particularly as regards the large and small forms of the plant.

Notes are highly desirable also as to whether Brent and Wigeon visit, in any numbers, places where *Zostera* has never been seen, as, for example, the Wash, or where *Zostera* is now very rare or extinct, as, for example, Lymington and Southampton. If, as is likely, they do visit such places, then the examination of the birds to discover what their food is on these marshes would be an important contribution to the solution of this problem.

Then there is the question of discrimination. Do birds pick out and feed on *Zostera* beds rather than on other equally suitable food plants that may be in the vicinity (see p. 43), and, as there are both large and small forms of the *Zostera*, do the birds show any preference for size or habitat? On the face of it, this seems very unlikely, except that they may resort to places where they can perceive the greatest quantity regardless of the size of the individual plants. Here again bird watching and stomach examination might answer the question.

Seasonal variation of the food must also be taken into account. As far as can be ascertained, *Zostera* remains growing in abundance throughout the year. Other plants, e.g. *Potamogeton pectinatus* (see p. 44), die down in the winter and, therefore, are not available during hard times. Similarly, in winter fresh-waters may freeze up, but the sea is almost always open.

Having ascertained the principal food plants, their natural distribution around our coasts and particularly whether *Zostera* is again on the increase and where, then, having gained an intimate knowledge of the habits of each species and factors which lead to its successful colonisation or to its extermination, one would be in a better position to consider what steps could be taken to make good any food-plant deficiency. The following notes on this problem are put forward in the hope that they will serve as a basis for future work.

Other Possible Maritime Food Plants

The question now arises as to what other plants grow where Wigeon and Brent might feed. The sea weeds in general appear to be of no importance and only one need be mentioned. That is the sea lettuce (*Ulva* and *Enteromorpha*). It is extremely abundant in a large number of the English estuaries as, for example, the Orwell and Southampton Water. Here, as in many other cases, it now covers the stones where *Zostera* used to grow. In appearance it consists of flat formless masses of green thallus, and as they are algae and not flowering plants no stem or leaf is formed.

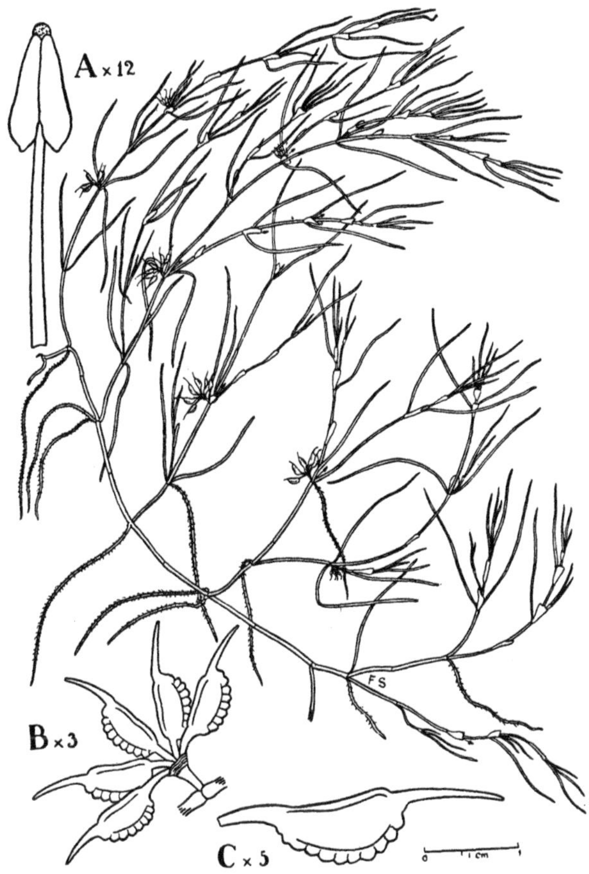

Fig. 5. Horned pond weed (*Zannichellia pedunculata*). *A*, stamen; *B* and *C*, fruits.

There are no other plants that grow on the open shore, but three closely related plants of some importance occur in lagoons and ditches above and between ordinary tidal limits. These are tassel pond weed (***Ruppia spiralis***), horned pond weed (***Zannichellia pedunculata***) and fine-leaved pond weed (***Potamogeton pectinatus***). All have much branched stems and numerous leaves about 3 inches long and not more than $\frac{1}{20}$th inch wide. They are distinguished by their flowers and seeds. *Ruppia* has long-stalked tassels of three to seven stalked seeds shaped rather like a duck; *Zannichellia* has bunches of boat-shaped seeds borne in rings at the base of the

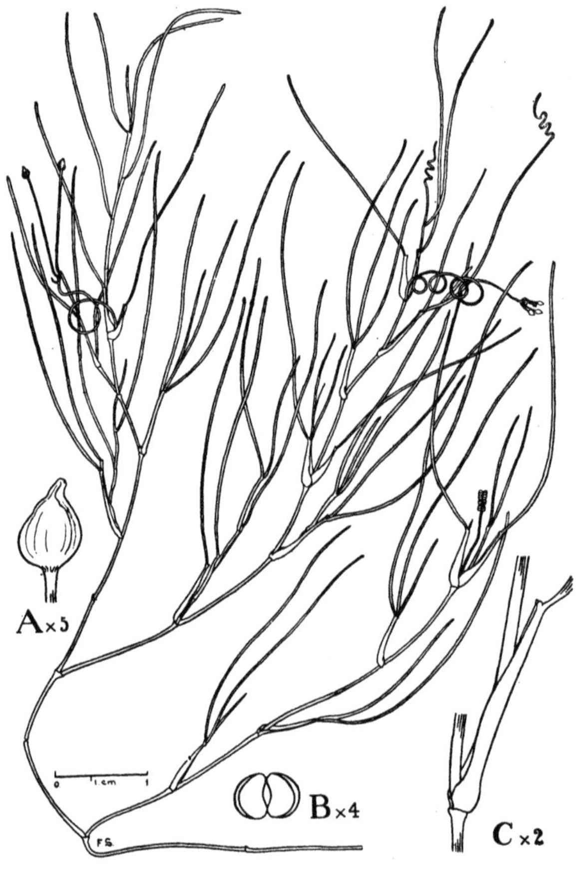

Fig. 6. Tassel pond weed (*Ruppia spiralis*). *A*, fruit; *B*, stamen; *C*, stipule.

leaves; *Potamogeton* has shortly stalked spikes of many oval seeds. All these plants are reddish green and are a marked contrast to the bright green of *Zostera*.

Ruppia can be found in quantity in lagoons of high salinity, higher in fact than in the sea. *Zannichellia* is usually seen in the dykes with limited flow and a salinity much lower than sea water, probably also below a strength suitable for *Zostera*. The *Potamogeton* is found in similar places and often in quite fresh-water. All these plants form part of the food supply of birds of all kinds, and probably Brent and Wigeon include them in their diet. It has already been noted that an important character about certain water plants is that their volume in winter is about equal to that found in the summer, and to this group *Zostera* belongs; *Ruppia, Zannichellia* and *Potamogeton* probably decrease rapidly in the autumn and, therefore, although valuable food in summer, may not be available in the same quantity in the winter.

Biological Conditions leading to Diminution of Food Plants, especially *Zostera*

There is no doubt about the decrease of *Zostera* in recent years, and four factors which may have contributed to this decrease are: change of substratum, disease, the devouring of the plant as a food by various birds, and competition with other plants. The first two have been discussed at length. In the relationship between an animal and its food it is important that the food be superabundant, and when such a condition obtains increase of food keeps pace with or exceeds the devouring of the food. If by any chance the food plant becomes scarce and amount of feeding remains the same, it is obvious that the equilibrium between growth and feeding will be upset and the feeders will tend to exterminate the food plant. This can easily be the final step in the denudation of *Zostera* beds already thinned by other causes. On the desertion of a denuded area by the feeders the plant will have a chance to recover if any reproductive portions are left.

In this matter, the feeding habits of the birds particularly under discussion may make a great difference. Wigeon, when not too much disturbed, feed on *Zostera* both night and day. They commence feeding as soon as the tide has ebbed sufficiently to allow them to reach the plant, and in dry weather they stop when the mud flats dry out completely. If very hungry they are known to feed even when the mud flats are dry. Their method usually is to nip off the grass blade by blade without waste.

Brent, similarly when undisturbed, feed by day and by moonlight on the mud flats. While feeding, they tear up a quantity of the plant, some of which they do not need at the moment. When the tide covers the flats, they drift off with the plant which they have torn up. When wind and tidal streams are suitable, they remain with the floating weed and continue to feed on this while the growing plant is inaccessible.

Swans when feeding tear up quantities of the grass with the rhizomes attached. Their long necks enable them to feed for a longer period over the tide than the other birds, and they do not supply themselves with a reserve of food in the form of detached and floating plants; it seems rather that they ignore to a great extent the plants they tear up, which is, therefore, wasted unless other birds take it. Because of this wasteful feeding habit of the Swan it is blamed for the shortage of *Zostera* in certain places. This surmise requires careful confirmation as, for years, a large flock of Swans has been kept at Abbotsbury, and until 1931 the quantity of food was sufficient.

Relation of *Spartina townsendii* to *Z. marina* on the South Coast

In 1836 a cord grass (*Spartina alterniflora*) was observed in Southampton Water and was said to have come from America. Up to then only a smaller species, *S. stricta*, had been recorded. Then in 1878 a large and more vigorous cord grass was discovered, also in Southampton Water, and this was called *S. townsendii* by Messrs H. and J. Groves in honour of Mr Townsend, an eminent botanist and worker on the Hampshire flora. This grass was more vigorous than either of the others mentioned, and it could, and did, grow on the bare mud flats where nothing else was seen. Such a plant was bound to be a successful colonist, and now it has spread and filled nearly all estuaries from Kent to Dorset. Because of its vigorous growth on bare saline mud flats it was found invaluable in helping to stabilise and reclaim such areas, and for this purpose it has been planted in Holland and in other English estuaries.

On the taxonomic side it has now been established that this grass is a true breeding hybrid that resulted from a cross between *S. stricta* and *S. alterniflora*, both of which are now extremely rare in the Southampton district.

It is inevitable that reclamation by this grass would change the outline of an estuary and one of the plants that suffers most from the change is *Z. marina*. After establishing itself, *Spartina* consolidates the surrounding mud with its matted roots and then collects more and more mud, gradually raising a mud bank first above water-level, except on high tides, and ultimately above high-water mark so that a dry bank is formed where mud was before. As a consequence the river channel is narrowed and the currents are increased. This results in a removal of sand and mud away from the seaward edge of the *Spartina* zone and a permanent bed becomes a shifting one, and it has been pointed out that shifting sandy beds are uninhabited. There seems to be no doubt that in this way much of the *Zostera* has been exterminated, not by the cord grass itself, but by the change the grass causes in the currents by its colonising activities. *Zostera* can sometimes be seen in among the *Spartina*, but subsequently the latter forms a closed community when nothing else can grow. So far, this extermination of *Zostera* by *Spartina* is only seen at Lymington and Southampton, but among other places it may also be going on in the Suffolk Stour and in Poole harbour.

Possible Methods of Conserving and Increasing the Food Plants

Whatever the cause, *Zostera* is clearly becoming scarcer around the coasts of Britain. In spite of many doubts and exceptions, migrating fowl, especially Wigeon and Brent, often feed on this plant, and their numbers are sometimes determined by the quantity of this plant available as food. Reports in some places show a recovery, in others the plant is still on the decrease, in yet a few places there is more now than previously. It probably makes no difference whether small- or large-leaved forms are available as food except as regards bulk of crop per acre, and for this *Z. marina* is to be favoured to *Z. marina* var. *angustifolia* or *Z. nana*.

Disease may be one factor, but it is also clear that often a change in the substratum, particularly from non-shifting to shifting, is the cause of the extermination of large quantities of eel grass. This suggests one method whereby the volume of *Zostera* might be increased, and that is by protection of mud and sand banks from scour, so that they remain non-shifting. The plant would undoubtedly flourish in artificially entrained lagoons, as it seems unnecessary for good growth that the tide should rise and fall over the beds, though it does so in many places. How land is to be made stable to grow sufficient quantities of *Zostera* to replace what has been destroyed is another of the problems that has arisen over this unprecedented decrease of the plant but, to the author, attention to the substratum seems the most promising line of investigation.

As regards alternative food plants there seems to be none except sea lettuce that could be cultivated on the open shore or on unprotected mud flats. In the salt marshes behind these, however, more pools and lagoons might be formed in which, if suitably protected, *Ruppia*, *Zannichellia* and *Potamogeton pectinatus* would grow; the first in very saline water and the other two in sub-saline or even fresh water.

Summary

1. Grass wrack, wigeon grass or eel grass includes three forms: *Z. marina* L. with leaves 6–4 mm. wide, *Z. marina* var. *angustifolia* with leaves 4–1 mm. and *Z. nana* with leaves 2–1 mm. wide.

2. At the present moment *Z. marina* is uncommon and apparently confined to sandy places near low-water mark and in certain lagoons. *Z. marina* var. *angustifolia* is the commonest and found in muddy estuaries and on the seashore. *Z. nana* is found in places similar to the last but chiefly in southern England.

3. *Zostera* has been decreasing in quantity for at least fifteen years, but there was an abnormally large decrease in 1931 and 1932. Since then, in some places, there has been recovery and in others a continuous decrease. *Z. marina* suffered most and *Z. marina* var. *angustifolia* least, but there are some exceptions.

4. The two most likely causes are a change in the nature and stability of the substratum and a decrease caused by a parasitic fungus. Mute Swans may do some damage in certain areas.

5. The relation of the quantity of *Zostera* and the birds that are known to feed extensively on it is discussed. More data are required as to the food of these birds, wherever they are known to collect in any quantity.

6. Alternative food plants and the possibility of increasing the crop of *Zostera* are discussed.

Acknowledgements

In the compiling of this report the author has received great help from the notes of Mr C. Dalgety and the County Recorders of the Wildfowl Committee.

References

(1) Atkins, W. R. G. *Journ. Marine Biol. Assoc.* xxiii, 207 (1938).
(2) Blackburn, K. B. *Nature*, cxxxiv, 738 (1934).
(3) Butcher, R. W. *Journ. Cons. int. Explor. Mer.* ix, 49 (1934).
(4) — *Rept. Bot. Ex. Club for* 1933, p. 592 (1934).
(5) Lynn, M. J. *Irish Nat. Journ.* vi, 107 (1936).
(6) Mounce, I. and Diehl, W. W. *Canad. Journ. Res.* xi (1934).
(7) Philip, G. *Journ. of Ecology*, xxiv, 205 (1936).
(8) Pond, W. H. *Rept. of U.S. Comm. Fish and Fisheries*, xix, 483 (1905).
(9) Renn, C. E. *Biol. Bull. Wood's Hole*, lxx (1936).
(10) Tutin, T. G. *Journ. of Botany*, lxxiv (1936).
(11) — *New Phytologist*, xxxvii, 50 (1938).
(12) Water Pollution Research Board. Technical Paper 7 (London 1938).

IV

BRITISH DECOYS

By H. A. GILBERT assisted by COLIN McLEAN,
C. W. MACKWORTH PRAED, H. LEYBORNE POPHAM

In Great Britain the decoying of wildfowl has been a declining industry for many years. Sir Ralph Payne-Gallwey, in his standard work on the subject, *The Book of Duck Decoys* (1886), gives a list of roughly 200 decoys which were then working or had been worked in past times in Great Britain, together with twenty-two more in Ireland. In 1815 when the famous wildfowler, Colonel Peter Hawker, visited the Fen country, he soon discovered that the numerous decoys prevented any sport with the gun. Under the date November 15th of that year he records in his diary: "Walked nearly 30 miles in surveying the Fens, and could soon perceive that they would not answer for wildfowl shooting; if a frost, the birds are gone; if a thaw, the greater part of them remain on the decoys." At that time no less than thirty-nine decoys were working in Lincolnshire alone, while nearly a hundred were in full use in the Eastern Counties. The decoying of wildfowl was an important industry, providing hundreds of thousands of duck for the market—important in the sense that no meat was imported in those days. The drainage of the Fens and the rapid increase of population caused this industry to decline, until in 1886 Sir Ralph Payne-Gallwey could discover only forty-four decoys in the whole of Great Britain and three in Ireland. By 1918 this number had been reduced still further. In that year Mr J. Whitaker (*British Duck Decoys of To-day* (1918)) found that only nineteen pipe decoys were in use. Eight of these are now disused:

Ivythorne (Somerset)	Hale (Lancs)
Oakley Park (Salop)	Ashby (Lincs)
Iken (Suffolk)	Thirkleby Park (Yorks)
Brooke Hall (Suffolk)	Hornby Castle (Yorks)

At the present moment (1936) four decoys only are working full time, and it is understood that some of these may close in the immediate future.

Decoys in Full Use (1936)

1. Borough Fen (Northants).
2. Fritton (Suffolk).
3. Orwell Park (Suffolk).
4. The Grange (Essex).
5. In addition another decoy, Orielton (Pembrokeshire), was reconditioned in 1934. It is now in full use but no duck are killed thereon. It is used for ringing purposes only.

Decoys partly in Use

		Average catch for last ten years
6.	Boarstall (Bucks). Used by the owner to supply himself and his friends	175
7.	Berkeley (Glos). Not used for the last five years but is now being reconditioned	150
8.	Abbotsbury (Dorset). Sparingly used chiefly for owner's and friends' requirements	390
9.	Marsh House (Essex). One pipe restored three years ago which is used during week-ends only (three years' average)	198
10.	Wretham (Norfolk). Two pipes in repair but rarely used, duck shot in flight	205
11.	Hamptworth (Wilts). Two pipes rarely used	30
	Total	1148

There have never been any decoys in Scotland, nor are there any in Northern Ireland. In the Irish Free State a law has been passed by which wild duck have been declared game birds, and the taking of all game birds by nets or traps has been declared illegal. Under this law the last decoys have been closed in the Irish Free State.

The number of duck caught on our decoys varies greatly from year to year. C. McLean, in his Presidential Address to the Norfolk and Norwich Naturalists' Society (March 1936), has recorded the numbers caught on three of our most important decoys, namely Borough Fen, Fritton and Orwell Park, during the last eleven years 1925–35 (see p. 52). These show an annual average of 8887 duck.

We must, however, draw attention to the fact that the numbers appear to have decreased greatly during the past three years, especially in the case of Wigeon. The cause of this decrease may not

BOROUGH FEN, FRITTON & ORWELL PARK

	Mallard	Teal	Wigeon	Other duck	Total
1925	10075	4336	1177	155	15,743
1926	2870	1916	2288	93	7,167
1927	5899	2159	1165	134	9,357
1928	4693	2744	2649	62	10,148
1929	2930	1404	3621	59	8,014
1930	2925	2466	2962	73	8,426
1931	3215	2845	3667	97	9,824
1932	3680	3349	1481	68	8,578
1933	4799	3266	896	220	9,182
1934	2995	3070	734	24	6,823
1935	2158	2092	239	6	4,495
				Total	97,757

THE GRANGE (ESSEX)

	Mallard	Teal	Wigeon	Total
1924	657	1329	1471	3457
1925	1575	1217	259	3051
1926	555	1036	525	2116
1927	1432	474	422	2328
1928	1338	590	247	2175
1929	740	959	305	2004
1930	730	618	17	1365
1931	703	441	481	1625
1932	666	462	619	1747
1933	The decoy was dry			
1934	116	237	11	364
1935	222	303	25	550
			Total	20,782

be permanent. It may be due solely to the climatic conditions which prevailed during the migration period or for some other temporary cause unknown.

The annual catch of wild duck in British decoys for the last ten years is, therefore, as follows:

Borough Fen, Fritton and Orwell Park	8887
The Grange	1732
The six smaller decoys	1148
Total	11,767

Dr P. R. Lowe in 1926 estimated the numbers at that time to be 14,000 to 15,000 more or less (*vide* the ***Present Status of the Wild Fowl of Europe***).

Effect of Decoys

Wild duck—subject as they are to incessant persecution by gunners—feed by night, and flock to a decoy by day because of the peace and quiet which is obtainable thereon. They are blissfully unaware of the slaughter which takes place amongst them in the quietude of secrecy.

In Britain the decoys which are worked for amusement only must be of great benefit to duck, since they give them daylight sanctuary in return for a very small toll of their numbers. Decoys, however, can be very destructive. At Orielton the numbers of duck frequenting the decoy were recorded from day to day during the season. These numbers varied greatly. Duck left the decoy never to return and new arrivals were incessantly visiting the decoy to take their places.

The catch (1350) for the season 1935 was found to be slightly in excess of the average numbers to be seen on the decoy by day, during the season.

Shooting over an abandoned decoy does not destroy many duck. In the case of Orielton, during the sixteen years when the decoy was not used, less than one-tenth of the number formerly killed on the decoy were obtained by the gun. When a decoy is shot over, however, the disturbance drives the duck elsewhere, and wherever they are driven, other gunners will destroy them. Hence it must not be supposed that because a decoy is abandoned, all the duck which used to be killed there, less the small proportion which are shot, survive. Wherever the birds are driven, although the gunners will probably not kill quite so large a number as were killed on the decoy, an appreciable number are certain to be shot.

Prohibition of Decoys

Decoys have been prohibited in several countries, e.g. Denmark and the Irish Free State. We are aware that there is an agitation in certain quarters in favour of the prohibition of decoys in this country. It would seem to us that our decoys are likely to die out without any external pressure being exerted on them, at any rate to die out as commercial undertakings.

We feel, however, that we must point out in this connection that there is a private property in decoys in law. The case (Keeble *v.* Hickeringill) is fully set out in his book by Sir R. Payne-Gallwey who quotes the judgment of Lord Chief Justice Holt (1706):

> "That the person who made the decoy should benefit by the skill, outlay and care requisite to construct and manage it."
>
> "That the owner of a decoy had every reason to expect the law to protect him from wilful injury to it."

Two hundred and thirty years later this decision was upheld by Mr Justice Macnaghten (Hollywood Silver Fox Farm, Ltd. *v.* Emmett, 1936 2 K.B. 468). In that case those who are interested can read the law as it now stands. It would appear that he who disturbs a decoy with malice, whether he stands on his own property or not, is in peril of the Common Law of England (see also Carrington *v.* Taylor 11 East 571).

Our existing decoys are all ancient, one having been worked by the members of the same family for over 200 years. If, therefore, Parliament ever passes an Act prohibiting decoys directly or indirectly, it would seem to be only fair that decoy owners should be compensated for the destruction of their private property, and for their capital outlay.

Foreign Decoys

It does not seem to us to be our duty as a British Sub-Committee to inquire fully into foreign decoys. Indeed it would be practically impossible for us to do so. Fr. Haverschmidt, in his monograph on Dutch decoys (*Ardea* xx), quotes Otto Leege (1907) as follows: "We would have learned the exact numbers of captured ducks, but the trapper remained discreetly silent on the point." Again Payne-Gallwey himself says that he has not made any study of Dutch decoys because that "feat would be a difficult task, for in Holland decoys are more or less shrouded in secrecy and their doings concealed from enquirers on the subject". A decoyman by the very nature of his craft is a silent, secretive and suspicious individual who "is always on the watch for intruders, and will on no account give them the slightest reliable information".

No foreigner could hope to obtain information as to the Dutch decoy trade—a powerful vested interest (*vide* Haverschmidt). Yet there are two points in the above monograph to which we think it is advisable to direct the attention of the Committee.

Firstly, we must refer to the position in North Brabant where there are at least twenty-five decoys, many of which "are in the hands of one man". The season in Holland generally closes on February 15th, but in North Brabant not until March 13th. The reason for this extension, to quote Haverschmidt, is because "when shooting ceases everywhere else and the supply is therefore small, high prices are obtained for ducks in London, the great market for Dutch caught birds". This significant statement concerns us and certainly merits the attention of our Committee—without comment on our part.

Secondly, though Haverschmidt states that he is unable to obtain "exact data about the number of decoys in Gelderland, South Holland and North Brabant", he gives the number of decoys in Holland as at least 145. He further estimates the average yearly captures at 2000 for every decoy—"a yearly total catch of 300,000 ducks in Holland". For the following reasons we are of opinion that the above is an under-estimate.

The upkeep and cost of a decoy is a heavy charge on an estate. One English landowner informs us that the maintenance of his decoy costs £220 per annum. The cost of another smaller decoy is known to be at least £130 a year. In the first case we are informed that the number of duck which must be caught before mere expenses can be met and before any profit can be made is 3500.

The balance sheet of a third decoy which is not worked for profit shows that the capture of each duck costs no less than five shillings.

In all these cases the payment of a decoyman's wages is included in the costs. It is possible that in Holland many decoys are worked by peasant owners and lessees as a part-time occupation; and that the working costs are proportionately less in that country.

We believe, however, that the average price obtainable for Mallard to be 1*s*. 9*d*. each; and for "half duck" (Wigeon, Teal, Pintail and Shoveler) to be between 9*d*. and 10*d*. each. Commission, freight and other charges must be deducted from these prices, with the result that the decoyman receives little more than 1*s*. a bird for his whole catch. Hence the numbers mentioned by Haverschmidt as the average catch per decoy in Holland—namely 2000 duck—would not pay even the expenses of any decoy in the world—much less yield a profit. Further in this connection we are informed that a tax of 1000 guilders is levied by the Dutch Government on every decoy, but since we have been unable to obtain confirmation of this fact we are compelled to leave it out of our calculations. One of our members, however, has been informed by a decoyman in Holland that a catch of over 3000 duck was necessary before the expenses of his decoy were recovered.

It would seem, therefore, that at least 500,000 duck must be killed annually on Dutch decoys before any profit can be made, and these decoys are not worked for amusement only. Every migration passes over Holland and the estuaries are large and numerous. Hence Dutch decoys are very favourably situated. They are known to have made profits and even more are believed to have made large profits. Though the retention of the gold standard by the Dutch Government has probably hit their export trade very hard during the last five years, what the annual slaughter of duck amounts to on the decoys of Holland is a question to which no man living—certainly no foreigner—is likely to obtain an answer.

We feel sure, however, that if 150 decoys are still working in Holland, the number of duck killed is largely in excess of Haverschmidt's estimate. This number certainly must be a million and probably amounts to 1,500,000. We express no opinion, however, as to whether this number is excessive. The Dutch have used decoys for centuries, and their decoys cannot be the prime cause for the decrease of wildfowl.

Conclusions

1. Decoys in Great Britain are rapidly dying out—at any rate as commercial undertakings.

2. The average number of duck caught in British decoys is 11,767 per annum, and these numbers are likely to decrease in future as more decoys go out of use.

V

PUNT-GUNNING

By C. T. DALGETY, assisted by R. A. H. COOMBES, H. A. GILBERT, H. LEYBORNE POPHAM, T. RUSSELL-GODDARD

The investigation into punt-gunning in England has proved difficult, chiefly from two causes. One, that punt-gunners have learnt by experience to mistrust the motives of any inquiry, and to expect a thoroughly one-sided report. The other, that many of the fisherman-fowler type are comparatively illiterate and do not keep any kind of written records.

However, many punt-gunners have been very generous in giving facts and figures to me and the other members of the group who have been carrying out the investigation. We are very grateful to those who have assisted us in this way. Amongst those who have helped are amateurs and "professionals", men who have spent a lifetime in one locality, and men who have visited various localities.

The aims of the investigation were to find out the following:

1. The number of punts working in England.
2. The number of professionals making part of their living by the sale of wildfowl.
3. The average and maximum bag per punt per season.
4. Bags of the present day compared with those of the past.
5. Average bag per day or night spent in pursuit of fowl.
6. The number of mechanically propelled boats carrying punt-guns.
7. Any causes of change in the number of wildfowl.
8. Suggestions for combating any decrease in numbers.

1. *The Number of Punts working in England*

It has not been possible to get in touch with every individual owner of a gunning-punt. A few may have escaped our notice completely, while others may have been counted twice over because of their moving from one place to another.

Some punters are men who live on the seashore and are prepared to go out on every possible occasion, while others are business men who go out for as little as one, or two, week-ends in the whole year.

It is certain that there are not less than 105 working punts, nor more than 150, in England.

For the purpose of arriving at these figures we have included all punts which mount a punt-gun and are used almost every season, although some of them may be used only once or twice in the season. So much work has been entailed in collecting the required information from England, that it has been impossible to do the same for Scotland. However it can be truthfully said that there are not less than twenty working punts in Scotland, and I doubt if there are as many as thirty.

2. *The Number of Professionals making Part of their Living by the Sale of Wildfowl*

Professional wildfowlers, in the true sense of the words, have almost ceased to exist. The standard of living is higher now, the present minimum agricultural wage is nearly three times the average agricultural earning of 100 years ago, and the public do not buy wildfowl to the same extent as they did, now that butchers' meat is cheaper. Also dead wildfowl are now imported in large quantities, which has a deteriorating influence on the prices obtainable for wildfowl killed in this country. All these factors combined have almost abolished the professional wildfowler, though there are about five true professionals still alive.

The "professional" wildfowler of the present day is either a guide to amateur fowlers or still continues to go wildfowling for the love of the sport, hoping to defray his expenses by the sale of the fowl which he kills. Before the green plover was protected in Great Britain there were a few more men who made a living by wildfowling, their chief bag being green plover for which there has always been a ready market.

3. *The Average and Maximum Bag per Punt per Season*

When one considers that there are so many different localities, and that different people spend amounts of time varying from three days to the whole season, it is apparent that it is impossible to arrive at a definite average bag for all the English punts; nor would such a figure be of much practical interest. However, the following figures have been extracted from the information received and give a good idea of the bags made.

The average annual bag for a punter who goes out regularly, taken from the bags of twenty-six punters over a series of years and from all English coasts, is 287 duck and geese.

The average annual bag for a good punter, working hard for the whole season in a good locality, is approximately 400 duck and geese.

The average annual bag made under ideal conditions by a good single-handed punter on a private estuary abroad in Europe was 700 duck and geese. It would be impossible to get such an average in England, because there are few estuaries so well suited and all of these are free for anyone to shoot over.

The largest bag recorded in any season during the present century, which was made by four men using the same double punt, was 1500 wildfowl. If only two men had used the punt the bag could not have been nearly so great. In fact the one punt being used by two different crews is almost the same as the use of two punts.

One single-handed punter, in the winter of 1917–18, killed over 1000 fowl. This occurred under very abnormal conditions, because all shooting in that area had previously been stopped owing to the war, and that winter was exceptionally suitable owing to the prolonged frost. The average annual bag of this same man is 152 fowl.

The average bag per annum per punt in Great Britain is estimated at between 100 and 150 duck and geese.

4. *Bags of the Present Day compared with those of the Past*

Unfortunately there is only one set of figures available for this comparison. The lack of written records and the ruination of wildfowl resorts by the "advance of civilisation" have made it impossible to secure other figures for comparison.

This one set of figures gives the results shown in Fig. 7. In each case the left-hand column represents the bags obtained during the first half of the nineteenth century, while the right-hand column represents those obtained during the present century.

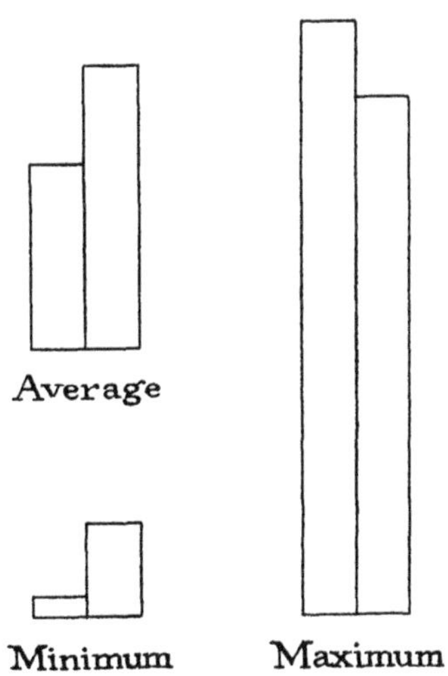

Fig. 7

The figures which have provided these results were taken from the "game book" of a sportsman who used his gunning-punt in one locality for thirty seasons during the first half of the nineteenth century, and compared with figures from the records of a punt-gunner who worked the same locality for eight seasons during the period 1927–34. It is interesting to note that the bags of the present day show an increase of 55 per cent on the bags of the past. The years 1927–34, which included some very bad wildfowl years, had no year as bad as the worst during the nineteenth-century period, nor did it have any year as good as the best during the nineteenth-century period. It seems probable that, in this particular locality, if the present-day records could be extended to cover a period of thirty years, there would be little difference between the present-day bags and those of the beginning of the last century.

5. *Average Bag per Day or Night spent in Pursuit of Fowl*

In bad seasons the average daily bag may be as low as one, while in good seasons it may be as high as fifteen. These figures are for those people who spend a reasonable proportion of the season in punt-gunning. Where a punter lives on the shore and only goes out when conditions are thoroughly favourable his daily average may be greater, although his season's average remains normal, or below average.

6. *The Number of Mechanically Propelled Boats carrying Punt-guns*

It is estimated that there are less than twelve power boats mounting punt-guns in England.

All true punt-gunners are unanimous in condemning the use of punt-guns mounted on power boats. Their reasons for this are not that they are destructive to wildfowl, but that they cause excessive disturbance and much wounding of fowl. The disturbance is caused by their attacking fowl which are resting on deep water at sea. Fowl have come to look upon deep water as a sanctuary. The breaking of this "sanctuary", if done persistently, drives the fowl away altogether. Almost every shot from a power boat is taken at birds on or leaving the water; this is necessitated by the deep draught of the boat. Under these conditions duck are nearly always in such a formation that it is impossible to shoot at them without the certainty of wounding the birds in the farther side of the flock.

7, 8. *The Causes of Change in the Number of Wildfowl, and Suggestions for Combating any Decrease in Numbers*

Much valuable information on these subjects was given by punt-gunners and has been included elsewhere.

Punt-gunning, as one of the many factors which influence the status of wildfowl, had to have a special inquiry. But there was further reason for this inquiry in the fact that many people, both British and foreign, are very much against punt-gunning. Hitherto there has been no unbiased report on punt-gunning, and consequently people without personal acquaintance with the sport have been unable to form any conclusions with regard to it. The present report is the result of the inquiry made by a group of the members of the Wildfowl Inquiry Committee. This group was composed of both punt-gunners and men with no addictions towards punt-gunning.

From the figures already mentioned it is possible to see the extent to which punt-gunning influences the status of wildfowl. Now we must consider the accusations brought by those who oppose the use of punt-guns.

Let us consider who are the people who would like to prohibit this sport. They can be divided into three sections. First, those who would like to prohibit all blood sports. Secondly, those who are jealous of the punt-gunner and think that if it were not for him they themselves would be able to shoot the same duck. Thirdly, those who consider the use of the punt-gun to be too destructive or too cruel. The first two sections are disregarded, but the third must have its accusations answered.

Briefly, the sins which they attribute to the punt-gunner are:

1. The punt-gunner kills more than his share of wildfowl.
2. He wounds a very large number of birds which die and are lost.
3. He causes too much disturbance of birds.
4. His methods are not consistent with the ideas of sport.

1. *Does the punt-gunner kill more than his share of wildfowl?*

At the end of the last century, Sir Ralph Payne-Gallwey wrote that anybody using a double-handed punt did extremely well to get 700 duck and geese in a season. That this statement holds good to-day is shown by the figures in this report: 700 duck and geese in one season is a very large bag for one double punt; for a single-handed punt it is an enormous bag. The average bag per punt throughout Great Britain is not one-quarter of the above figure.

Let us consider that 700 duck and geese is a large bag which it is possible to kill, but very unlikely to exceed, with a double punt in one season. That means 350 duck and geese to each of the two men. This is not a disproportionate quantity for one man. Yet this number represents a far bigger bag than usual, in fact a bag which was never attained by the famous Colonel Peter Hawker.

People have conceived the idea that punts are more destructive because they hear of the "big shots" and do not hear of the "small shots", also they have no idea of the number of shots that are fired in a season or a day. The number of days, or sometimes almost whole seasons, which are unsuitable is another thing which people do not take into consideration.

The following figures are a fair sample for a good season. A pair of punt-gunners were prepared to go out on 100 days. The tides and weather conditions allowed them to go out on only forty-five of those days. Their bag averaged thirteen birds per day out. They fired an average of less than one and a quarter shots per day out. They killed an average of eleven birds per shot.

I have found that these figures seem incredibly small to those who have criticised punt-gunning; yet I can vouch for the fact that they represent a bag of very much above the average, in fact a bag made in an outstandingly good season.

The bags made by so-called "sportsmen" who took their punts to the Nile in Egypt, etc., have no parallel in Britain.

2. *Does the punt-gunner wound a large proportion of birds?*

I am absolutely certain that a far smaller proportion of wounded birds are lost by punt-gunners than by flight shooters. In daylight very few wounded birds are lost by good punters, while at night shots are fired at such close range that practically every bird is killed outright. It is impossible to prove these statements because there can be no record of lost birds, but the fact that a greater weight of shot is fired off per bird by flight shooters than by punt-gunners goes some way towards showing it. But mistakes do occur, especially with novices, when excessive wounding is caused by shooting at too great a range. However, this abuse is not confined to punt-guns.

3. *Does the punt-gunner cause too much disturbance of birds?*

One aim of all good punters is to cause the minimum possible disturbance, because disturbance spoils his sport. The discharge of a large punt-gun undoubtedly causes considerably more disturbance than the discharge of a 12-bore; but the punt-gun is fired less than twice a day and by no means every day, while the 12-bore is fired a great number of times in a day and every day.

4. *Are the punt-gunner's methods unsporting?*

It is impossible to compare one sport with another because the preference for one rather than another depends on individual taste. Briefly, a punter brings into play his own wits and skill to enable him to work his punt to within shot (i.e. between 100 yards and 60 yards) of a company of fowl, timing his stalk so that he will arrive within range at the moment most suitable for a successful shot. Having got within range, he must judge the exact moment when to shoot and must shoot straight.

When one considers that the punter fires an average of less than five shots in four days out, I do not think that anyone can call his sport an easy one. He must have extensive knowledge of the habits of wildfowl, and also of every other species of bird which lives on his "ground"; in fact he must be an accomplished field naturalist. He must be something of a seaman and able to handle his boat under all conditions. He must be physically fit in order to be able to manage his boat. In the management of his gun he must have very quick, or instantaneous, judgment, and must be able to shoot straight, estimate distances and estimate the "lead" required for a shot at a flying target.

If the sport were "unsporting" it would not have been among the favourite sports of thoughtful people with artistic, literary and musical talents, such as Colonel Peter Hawker, Abel Chapman, J. G. Millais and Frank Southgate. Abel Chapman did a great deal towards the preservation of the wild fauna of Europe and Africa, yet his favourite sport was punt-gunning.*

* Abel Chapman was the prime mover in the making of the Sabi Game Reserve, which later became the Kruger National Park, in the Transvaal. He used his influence to gain the much-needed protection for the Spanish ibex, and to attempt the same for the Norwegian reindeer.

VI

RINGING OF DUCK AT BRITISH DECOYS FOR THE WILDFOWL COMMITTEE

INTRODUCTORY

Of paramount importance for the work of the Wildfowl Committee is the ringing of duck with a view to tracing their migrations to and from their breeding areas in spring and autumn. For instance, if definite knowledge is obtained of the breeding areas of such ducks as winter in the British Isles, or pass through them on migration, it may be possible to exert influence in the matter of protection in those areas. Ducks are, fortunately, the easiest kind of birds to study in this manner: owing to the number shot and handled annually, the proportion of recovery records is high and the work of ringing is thus notably fruitful. Much valuable information about the migration of ducks has indeed already accrued from ringing activities in various countries, and a general survey of the results—with special reference to the British Isles—is given in chapter VII.

A great deal nevertheless remains to be learnt from an extended use of the method, and the Wildfowl Committee have accordingly fostered a scheme for ringing duck on a large scale, particularly birds caught in winter at decoys. The rings now used are those issued by the Bird-Ringing Committee of the British Trust for Ornithology, and bear the address of the British Museum (Natural History). The work is in progress at four decoys in Great Britain—Orielton, Pembrokeshire; Abbotsbury, Dorset; Borough Fen, Northamptonshire; and Somerleyton, Suffolk. An account of what has already been done at each of these places is given in the succeeding sections, leaving Orielton to the last—although first in the field—because it provides the largest amount of material.

A lively interest has also been taken in similar work in other parts of Europe, especially at decoys in Holland and Germany. A decoy in Denmark, too, provided many records at an earlier period. Much ringing of duck is now also being done in south-eastern Russia, where the birds are caught in late summer while undergoing wing-moult. Ringing of ducks in the breeding season, mostly as young birds, has taken place on a considerable scale in Iceland, Sweden, Finland and elsewhere. Reference to the results of this work abroad is made in the next chapter.

ABBOTSBURY DECOY, DORSET

Lord Ilchester is assisting the investigations by ringing and releasing a proportion of the duck caught in his decoy at Abbotsbury. The decoy, which was established in 1656, has three pipes, and during the last ten years the average catch has been 309 a year. During this time the decoy has been, and is continuing to be, used sparingly, chiefly for the owner's and his friends' requirements.

During the season 1937–8 the total catch amounted to 199 Mallard, 256 Teal and a few others, of which the following were ringed and released during the months November to February:

Drake Mallard	6
Duck Mallard	23
Drake Teal	50
Duck Teal	47
Drake Wigeon	1
Drake Pintail	1

Up to date there have been no recoveries of Mallard, with the exception of one duck Mallard found dead in the vicinity of the decoy.

There have been three recoveries of Teal within the British Isles, one Teal ringed at Abbotsbury in November 1937 being caught again at Orielton decoy on January 6th, 1938.

The following are the recoveries of Abbotsbury Teal from abroad:

Ringed	*Recovered*	
15. 2. 38	Shot, Klein Barkau, Preetz, Schleswig Holstein	22. 9. 38
15. 2. 38	Shot, Kuortane, Vaasa, West Finland	15. 8. 38
15. 2. 38	Shot, Nissum Fjord, Jylland, Denmark	31. 8. 38
20. 1. 38	Shot, Ijsselmeer, nr. Kampen, Holland	27. 8. 38
13. 1. 38	Shot, Enkoping, Sweden	25. 9. 38
15. 2. 38	Hamburg, Germany	11. 10. 38

BOROUGH FEN DECOY, NORTHAMPTONSHIRE

This decoy, which is situated a few miles from Peterborough, is one of the oldest in existence; all five pipes are still worked. The average catch over the last twelve years has been rather over 2000, chiefly Mallard, Teal and Wigeon. During the season 1937–8, 4300 duck were taken: Mallard, 1392; Teal, 2566; Wigeon, 324; Shoveller, 17; Pintail, 1.

In order to assist the Wildfowl Inquiry, 67 Wigeon were purchased by Mr Colin McLean in January 1938, taken to Dereham fifty miles distant, ringed and released. These consisted of twenty drakes and forty-seven ducks—taken as caught. Four recoveries of these ringed Wigeon were reported from the British Isles in 1938, and there was one particularly interesting recovery from abroad:

Ringed	*Recovered*	
26. 1. 38	Comacchio, nr. Ferrara, Emilia, Italy	15. 10. 38

SOMERLEYTON DECOY, SUFFOLK

This decoy, which is situated near Lowestoft, had originally several pipes, of which only two pipes at the edge of Fritton Broad are now worked. The average catch over the last twelve years has been about 1200, chiefly Mallard, Teal and Wigeon. During the season 1937–8 the catch was 2170: Mallard, 1713; Teal, 160; Wigeon, 272; Pintail, 7; Tufted 16; Pochard 2.

During January 1938 the following birds caught at the decoy were purchased by Mr Colin McLean, ringed and released:

Drake Wigeon	25
Duck Wigeon	56
Duck Pintail	4
Drake Pochard	2
Duck Pochard	3
Drake Tufted	4
Duck Tufted	6

Of these birds there have been ten recoveries of Wigeon, one of Pochard, and one of Tufted in the British Isles, in addition to three Tufted recaptured in the same decoy.

There have been three recoveries of Wigeon abroad as follows:

Ringed	*Recovered*	
28. 1. 38	Shot at Heydekrug, Memelland	9. 10. 38
14. 1. 38	Shot at Tiegenort, Danzig	22. 9. 38
14. 1. 38	Shot at Le Havre, Seine Inf., France	18. 12. 38

(A Teal ringed for the Leyden Museum at Makkum, Friesland, on December 15th, 1937, was taken at Somerleyton in January 1938.)

ORIELTON DECOY, PEMBROKESHIRE

By C. W. MACKWORTH PRAED

In 1934, two members of the present Wildfowl Inquiry Committee reconditioned an ancient duck decoy. This was situated at Orielton in Pembrokeshire, and, as it was the former home of one of them, the old records were available, as well as some local knowledge of the past. The pond had been shot over regularly for some seasons and the "pipes" were choked with young trees and silted up with mud, so it was with great interest that the gradual return of the fowl was witnessed.

As we wrote at the time—when we first saw Orielton on a February day, the old decoy presented a dismal sight. Could it ever be restored from chaos to tidiness? Even if we refashioned the lost labour of past generations, should we obtain results commensurate with the trouble that we must take, and the difficulties that must be overcome? Yet there was one slight gleam of hope. Overhead round and round above us flew a nice bunch of Teal, a few Wigeon, a dozen Shoveler, and a flock of Tufted Duck. The birds never came within shot, but refused to leave the old decoy pond. Even after twenty years of shooting they were obviously attached to the place. They were, in truth, a mere remnant of the thousands which years ago had made the decoy their daylight sanctuary, but at any rate they were a useful "lead in". Given peace and quiet they must bring others to the decoy after every nightly forage for food.

On that slight encouragement, and against the advice of all the wise, we decided to rebuild the old decoy again from the beginning. There was a chance that we might learn something about the migrations of duck.

In the old days, the average daily population of the lake at the height of the season was probably 10,000–12,000 duck, of which some 90 per cent. were Wigeon. The big flocks of Wigeon had, however, long left the decoy, and even in the estuaries nearby their total number was not more than 3000. Teal, on the other hand, were always present, and even in our second year we took a census of Teal which made out not less than 2300 birds on the pond, and that was probably as many as there ever had been. Wigeon have returned slowly, at first mainly at night, but only after three years have they shown up in respectable numbers, and this year there were for a short time some 3000 birds on the lake. There is also, and always was in the old days, a flock of some hundreds which graze all day in a densely packed mass on the grass of the adjoining meadow. This is a habit they have always had, and it is noticeable that each fresh flock seems to acquire this habit, and in a very short time leave the water and forage on the meadow quite a long way from the lake.

Whatever may be the cause, the present population of duck is very different from what it was in the days before the 1914 war. Then, Pintail and Shoveler were comparatively rare; now 100 pairs of Shoveler, and half as many Pintail, is a very normal count. On the other hand, there are far fewer diving ducks, especially Pochard: there are probably not more than 30 Pochard on the lake, and there used to be dozens if not hundreds. Mallard were never numerous, nor are they now, but on balance they have probably decreased over the last twenty years. In short, the only really serious diminution is in the number of Wigeon, but this is very great indeed.

It has been most interesting to watch the effect that quiet and preservation have had on the duck, and we now turn to the actual work in ringing done in the last few years, latterly on behalf of the Wildfowl Inquiry Committee. The actual numbers ringed have been

1934–5	1935–6	1936–7	1937–8	1938–9 (to 4 Jan.)
326	1259	1176	1432	1577

The number of recoveries—not counting recaptures—have been in all 506 to date, of which 171 have been from overseas as follows: Ireland, 56; Scandinavia, 32; Russia, 14; France and Belgium, 15; Holland, 31; Germany, 14; Latvia, 3; Italy, 2; Iceland, Spain, Poland and Rumania, 1 each.

The number of recaptures at Orielton is very large indeed, 800–1000 being normal in a season, and while they entail an enormous amount of work they are also very interesting. One Teal, original number 66, has been recaptured for the fifth season running.

The actual species and number of duck that have been ringed in the last few years are as follows:

Teal	3700	Tufted	100
Wigeon	370	Pochard	3
Mallard	260	Gadwall	1
Shoveler	170	Garganey	1
Pintail	115	Smew	1

It will be seen, therefore, that the only species we can talk about with some certainty is the Teal, and we can make slight conjectures at the movements of Wigeon and Shoveler. Mallard are so far uninteresting, nearly all our records being local.

Teal unquestionably pass on west from us in the late autumn—if they pass on at all—and our main volume of winter records of autumn-ringed birds is from Ireland. About January, when the floods are out, the Teal spread eastwards, and we get records in January and February from all over the British Isles, and in February and March from both sides of the Channel. This becomes more pronounced in March and April, and it seems that, unlikely as it would appear, the main migration route of our birds is first south and then up-Channel. It need not be stressed that this is the movement from which the Dutch duck decoys derive a considerable advantage, and it is one more reason for curtailing the season for the importation of dead birds into this country.

As regards arrival, we seem to get two main waves of both Teal and Wigeon, the first in late October to mid-November, and the second—unusually pronounced in 1938 owing to hard weather—about Christmas. These latter birds are often in bad condition, whereas the earlier migrants are as fat as butter.

A very great majority of our birds turn up in a second season, as our recaptures show, but we also get a fair number of recoveries in second and subsequent seasons which are very far afield. This is possibly due to the pairing of a bird from our winter area with another from a totally different one and then to our bird accompanying its mate to the latter's winter quarters: this probably explains our recoveries from Italy, Rumania, etc.

As it is much easier to trace the movements of migrants from a map, we propose to reduce the letterpress to a minimum. The maps we give are as follows:

Fig. 8. TEAL: Recoveries in October–December of same season after autumn ringing at Orielton Decoy.

Fig. 9. TEAL: Recoveries in January–February of same season after autumn or winter ringing at Orielton Decoy.

Fig. 10. Teal: Recoveries on spring and autumn migration of birds ringed at Orielton Decoy. (A dozen recoveries in Holland and a few in the Frisian Islands are not separately shown.)

Fig. 11. Teal: Recoveries in summer (May–August) of birds ringed at Orielton Decoy.

Fig. 12. Teal: Recoveries in second and subsequent winters (some probably still on migration) of birds ringed at Orielton Decoy. (102 birds recovered at or near the decoy are not shown.)

Fig. 13. Wigeon: Recoveries in summer or early autumn of birds ringed at Orielton Decoy.

Fig. 14. Pintail, Shoveler and Tufted Duck: Recoveries in summer or autumn of birds ringed at Orielton Decoy.

The actual reports, as may be seen from the maps, would by now make a formidable volume, but as they reach us—in all languages, but largely in excellent English—they are of great interest in themselves. They vary from the strictly official to the somewhat informal, and from two lines to four pages in length. Here it is possible to quote only the essential particulars of birds recovered abroad, and these are given in the appended table (pp. 72–76).

Of the recoveries at home and abroad, no less than 436 refer to Teal, and 10 to Shoveler, 26 to Wigeon, 16 to Mallard, 7 to Pintail and 2 to Tufted Duck.

A secondary interest is the ringing and releasing at a distance of waterfowl. Coots and Moorhens are an unmitigated nuisance on a decoy, as both, if in the "pipe" at the time of a catch, fly back past the decoyman and lead the duck out. We find, however, that we have to take them some distance away, as Moorhens return freely from 12 to 15 miles and Coots have returned from over 50 miles.

Other birds get into the pipes or traps from time to time, and we have ringed Snipe, Herons, Cormorants, Kingfishers, Woodcock, Dabchicks, Redshanks and Water-Rails, among others. In the old days Peregrines were commonly caught, but so far, although they are just as numerous, none have followed the duck into the pipes while the decoyman was present.

The majority of our duck are caught either by showing the dog or by feeding into pipes in the usual manner, but we have a few refinements when they know the dog too well. The stuffed forequarters of a fox, with a moveable head, is a considerable draw—as is also just a whisk past the screen of a fox's brush—while a ferret can also be employed occasionally. In addition, we have a certain number of cage traps, which are visited and cleared after dark, and it is in these last that the majority of our diving ducks are caught.

The proportion of sexes of the duck caught is remarkably constant and reveals an interesting state of affairs, as it always consists of considerably more females than males. This is true also if a census of the birds on the pond is taken, so it is not due to any difference in mental characteristics between the sexes. The proportion is so striking that we give figures in full as follows:

Sexes of duck caught

	Wigeon		Teal		Shoveler		Pintail		Tufted	
	M.	F.	M.	F.	M.	F.	M.	F.	M.	F.
1934–5	2	2	41	63	—	—	—	—	7	6
1935–6	24	41	231	321	11	25	—	—	4	12
1936–7	22	55	318	553	4	5	17	30	1	3
1937–8	8	22	354	732	16	14	5	16	26	29
1938–9 to date	56	132	401	706	38	53	25	20	7	11
Total	112	252	1345	2375	69	97	47	66	45	61

The mate of the Pintail recorded from Iceland returned to the decoy and was recaught this season. Five Garganey visited the decoy in July—of which one was caught and ringed. Two Teal ringed together were reported from about 50 miles apart in northern Sweden. Two Teal returned across the North Sea to Holland in early January. They are not shown on our maps—and one Teal was shot in Manche, France, 10 days after being ringed in December.

One Teal had been caught by us so often that when it entered a fish trap in Lapland it was so tame that the fisherman took it home and tried to keep it as a pet. One Teal caught in November 1935 and sent to the Zoo in London, where its flight feathers were clipped, turned up again in Pembrokeshire in December 1938 and was shot near the decoy. A second of the same batch was shot at King's Lynn in December 1937.

Neither of us as yet have shot one of our own ringed duck in other parts of England, but Captain Gilbert was out with a party in Herefordshire when one was shot, and they have been secured both above and below me on the Avon in Hampshire. A curious coincidence was the shooting of two of our Teal by two brothers, one at Euston in Suffolk and one at Quy in Cambridgeshire.

The reactions of duck to a dog are interesting. It appears to be almost entirely curiosity, with perhaps some tendency to mob—very like a herd of young cattle. Duck soon get used to a dog and evince little interest—in fact we can at once tell when a new lot of fowl are in by the amount of fuss they make when a dog appears. We used in the old days to put a red coat on the dog at times, to make some variety in the performance. Teal are the most inquisitive duck, then Mallard and Wigeon. The sense of smell of all surface-feeding duck is acute, but a burning piece of peat held in front of the mouth kills the smell of man completely: the difference can be demonstrated any day.

It is interesting to see what frightens duck. Aeroplanes nowadays have little effect compared with what they had even five years ago. Peregrines they always fear, and leave the land for the water at once. Buzzards they take not much notice of, but they dislike Great Black-backed Gulls very much. An otter on the pond is a devastation and the duck will never rest while one is about. One curious instance was that in which a sheep strayed into the pipe and put his head through the hole in the screen. The duck appeared to be so frightened that they avoided that particular spot for days!

As to the weather, we can say little, except that duck do not come in from the sea in any numbers if there is a fog; and they do not like the pond when it is flooded and always stay away to a considerable extent. They are much more difficult to catch in the rain, as the drip from the wire netting at the top of the pipe warns them that they are in a covered-in area.

So far, we have recovered very few birds ringed elsewhere—one from Abbotsbury and two from Piaam in Holland are all that we have come across—but it is to be hoped that considerably more ringing will be done in the near future, and that we shall find ourselves of use to other ringing stations in giving information of the movements of their birds.

RECOVERIES OUTSIDE BRITISH ISLES

TEAL

Recovered May

Ringed		
11. 11. 37	Shot S.E. of Lake Ilmen, RUSSIA (58° 15′ N., 31° 30′ E.).	3. 5. 38
9. 11. 37	River Cevernaya, Dvina, near Fedkovskoye, prov. Archangelsk, RUSSIA (64° N., 41° 50′ E.).	15. 5. 38
27. 11. 35	Shot, Lake Ilmen, 100 miles S. of Leningrad, RUSSIA (58° 15′ N., 31° 20′ E.).	16. 5. 36
20. 11. 37	Pechora river, near Rosvinskoye, RUSSIA (66° 33′ N., 52° 29′ E.).	29. 5. 38
25. 12. 35	Caught in fishing net, Kuhmojarvi, FINLAND (64° 11′ N., 30° 15′ E.).	14. 5. 37

Ringed		
27. 11. 36	"Found dead in a wire cage set for fishes", Vesanka, FINLAND (62° 15′ N., 25° 16′ E.).	26. 5. 37
24. 9. 36	Caught in fish trap, Gargnas, Västerbotten, Lapland, SWEDEN.	3. 5. 37
30. 1. 37	Shot 25 km. N. of Umea, Västerbotten, SWEDEN (63° 45′ N., 20° 20′ E.).	6. 5. 37
12. 2. 36	Starkesmark, Tvaraback, Västerbotten, SWEDEN.	— 5. 36
17. 1. 36	Killed by cat, Ofvanaker, SWEDEN (61° 20′ N., 15° 56′ E.).	— 5. 38
22. 11. 35	Shot, Tromsö Fylke, NORWAY (69° 35′ N., 19° E.).	28. 5. 36

Recovered June

8. 1. 38	Upper river Indiga, Lake Lebyashye, RUSSIA (67° 18′ N., 49° 55′ E.).	5. 6. 38
18. 1. 36	Dead, Lake Horn Ajvan, Lapland, SWEDEN (66° 15′ N., 17° 40′ E.).	— 6. 36

Recovered July

24. 11. 36	Shot, Strutele, near Tukums, LATVIA (56° 47′ N., 40° 36′ E.).	25. 7. 37
1. 12. 35	Shot, Rummelsburg, Pomerania, GERMANY (54° N., 16° 59′ E.).	19. 7. 36
12. 2. 36	Shot, Appelscha, S.E. point of Friesland prov., HOLLAND (52° 55′ N., 6° 40′ E.).	27. 7. 38

Recovered August

21. 12. 36	Shot, Salla, FINLAND (67° N., 29° 10′ E.).	20. 8. 37
23. 1. 37	Shot, Kittila, Lapland, FINLAND (67° 30′ N., 25° E.).	21. 8. 37
27. 12. 36	Shot, Munsala, W. FINLAND.	23. 8. 38
14. 11. 36	Shot, Lastringe, 23 km. N. of Umea, Västerbotten, SWEDEN (63° 45′ N., 20° 20′ E.).	4. 8. 37
15. 2. 37	Shot, Stora Lundby, Göteborg, SWEDEN (57° 41′ N., 11° 57′ E.).	19. 8. 38
2. 12. 37	Shot, Lulea River, SWEDEN (65° 35′ N., 22° 10′ E.).	26. 8. 38
19. 12. 35	Shot, Hoburgen, Gotland, SWEDEN (56° 54′ N., 18° 9′ E.).	29. 8. 36
8. 11. 35	Shot, Kuryly, Sokolka, Bialystok, POLAND (53° 24′ N., 23° 28′ E.).	18. 8. 36
18. 1. 38	Shot, Randersfjord, Dylband, DENMARK (56° 10′ N., 9° E.).	17. 8. 38
7. 11. 37	Aferso, Smaalands Havet, Skelskon, South Zeeland, DENMARK.	28. 8. 38
26. 10. 35	Shot, Island of Gjol, Limfjorden, DENMARK (57° 5′ N., 9° 43′ E.).	— 8. 38
20. 12. 35	Shot, Roskilde Fjord, DENMARK (55° 50′ N., 12° E.).	— 8. 37
7. 12. 35	Shot, Marienfelde, Deutsch Krone, Grenzmark Posen, GERMANY (53° 16′ N., 16° 28′ E.).	3. 8. 37
17. 1. 36	Shot, Himmelmoor, Quickborn, Holstein, GERMANY (53° 45′ N., 9° 55′ E.).	5. 8. 37
25. 1. 35	Shot, Beilen, Drenthe, HOLLAND (52° 52′ N., 6° 31′ E.).	5. 8. 36
27. 12. 37	Shot, Béthencourt-sur-Somme, FRANCE (49° 56′ N., 2° 55′ E.).	14. 8. 38
14. 10. 36	Shot by an "huttier", Conde-kerque, 5 km. from Dunkerque, FRANCE (51° 2′ N., 2° 22′ E.).	15. 8. 37
27. 11. 37	Shot, Palluel Marsh, S. of Lille, N. FRANCE (50° 38′ N., 3° 4′ E.).	21. 8. 38
28. 11. 36	Lake Ciair, Turlachi, Cetatea Alba Bessarabia, RUMANIA (46° 13′ N., 30° 17′ E.).	22. 8. 38

Recovered September

Ringed		
30. 1. 37	Shot, Lake Skattrask, Arvidsjaur, SWEDEN (65° 35′ N., 19° 10′ E.).	1. 9. 37
11. 12. 35	Shot, Bjorken, Ställdalen, Orebro, SWEDEN (59° 54′ N., 14° 57′ E.).	14. 9. 36
15. 1. 37	Shot, Nesbyen, Hallingdal, NORWAY (60° 30′ N., 9° E.).	— 9. 37
11. 2. 36	Lögstör, Limfjorden, Jutland, DENMARK (56° 58′ N., 9° 17′ E.).	26. 9. 38
24. 11. 37	Island of Moen, Zeeland, DENMARK (54° 57′ N., 12° 20′ E.).	— 9. 38
2. 12. 37	Shot near Waren, Muritz-See, Mecklenburg, GERMANY (53° 25′ N., 12° 40′ E.).	5. 9. 38
15. 1. 38	Shot, Ückermunde, Pomerania, GERMANY (53° 45′ N., 14° 10′ E.).	18. 9. 38
23. 12. 35	Caught in decoy, Island of Pellwurm, GERMANY (54° 32′ N., 8° 40′ E.).	19. 9. 37
29. 12. 37	Lohne, Oldenburg, GERMANY (53° 9′ N., 8° 13′ E.).	30. 9. 38
18. 11. 37	Caught in decoy, Föhr, North Frisian Islands, GERMANY (54° 42′ N., 8° 30′ E.).	30. 9. 38
4. 2. 37	Shot, Moerdijk, HOLLAND (51° 42′ N., 4° 38′ E.).	3. 9. 37
29. 12. 36	Caught, Makkum decoy, Friesland, HOLLAND (53° 4′ N., 5° 24′ E.).	4. 9. 37
20. 12. 35	Caught in decoy, Niverdyk, 10 km. from Dordrecht, HOLLAND (51° 43′ N., 4° 39′ E.).	5. 9. 36
14. 11. 36	Caught in decoy, Piaam, Friesland, HOLLAND (53° 2′ N., 5° 30′ E.).	16. 9. 38
25. 12. 36	Caught in decoy, Piaam, Friesland, HOLLAND (53° 2′ N., 5° 30′ E.).	16. 9. 38
24. 10. 36	Shot, Overyssel, HOLLAND (52° 25′ N., 6° 30′ E.).	16. 9. 37
10. 11. 37	Dordrecht, HOLLAND (51° 48′ N., 4° 39′ E.).	23. 9. 38
8. 11. 37	Caught in decoy, Piaam, Friesland, HOLLAND (53° 2′ N., 5° 30′ E.).	23. 9. 38
11. 9. 36	Killed, Makkum decoy, Friesland, HOLLAND (53° 4′ N., 5° 24′ E.).	24. 9. 37
26. 12. 35	Killed, Makkum decoy, Friesland, HOLLAND (53° 4′ N., 5° 24′ E.).	28. 9. 36
28. 12. 35	Killed, Makkum decoy, Friesland, HOLLAND (53° 4′ N., 5° 24′ E.).	28. 9. 36
24. 11. 36	Sent by A. S. Juniper and Co., Smithfield Market. Found in consignment from Workum, HOLLAND (port of export from Makkum decoy).	— 9. 37
11. 1. 37	Shot by an "huttier", Merlimont, near Etaples, Pas de Calais, FRANCE (50° 27′ N., 1° 37′ E.).	27. 9. 37
5. 11. 36	Shot, St Etienne d'Orthe, Landes, FRANCE (44° N., 0° 30′ W.).	10. 9. 37

Recovered October

25. 12. 35	Found dead, Holmarna Island, Nykarleby, FINLAND (63° 32′ N., 22° 45′ E.).	8. 10. 36
6. 1. 38	Odsmal, near Solberga, Bohns, SWEDEN (57° 51′ N., 12° 2′ E.).	5. 10. 38
29. 12. 35	Shot, Wase, near Filipstad, SWEDEN (59° 41′ N., 14° 10′ E.).	12. 10. 36
19. 12. 36	Falkoping, SWEDEN (58° 9′ N., 13° 33′ E.).	18. 10. 38
23. 11. 37	Shot, Aaras, Austrheim, Hordaland NORWAY (60° 20′ N., 4° 30′ E.).	12. 10. 38

Ringed		
13. 2. 36	Shot, Tating, Kreis Eiderstedt, near Vollerwick, Schleswig, GERMANY (54° 18′ N., 8° 43′ E.).	4. 10. 36
1. 2. 38	Island of Föhr, GERMANY (54° 42′ N., 8° 30′ E.).	10. 10. 38
17. 10. 36	Shot, Lehe Munden, Holstein, GERMANY.	18. 10. 37
13. 11. 35	Caught in decoy on Föhr, North Frisian Islands, GERMANY (54° 42′ N., 8° 30′ E.).	28. 10. 36
2. 1. 37	Zoutkamp, Groningen, HOLLAND (53° 13′ N., 6° 34′ E.).	3. 10. 38
28. 12. 37	Piaam, HOLLAND (53° 2′ N., 5° 30′ E.).	20. 10. 38
	Recovered November	
10. 11. 37	Saltholm, Oresund, DENMARK (55° 40′ N., 12° 47′ E.).	6. 11. 38
1. 1. 36	Caught in Boldicumer decoy, Föhr, GERMANY (54° 42′ N., 8° 30′ E.).	1. 11. 36
13. 2. 36	Neufeld an dem Elbe, Hamburg, GERMANY (53° 33′ N., 10° E.).	18. 11. 38
2. 12. 37	Neufeld an dem Elbe, Hamburg, GERMANY (53° 33′ N., 10° E.).	18. 11. 38
14. 11. 35	Caught in decoy near Hook of HOLLAND (51° 59′ N., 4° 7′ E.).	18. 11. 36
12. 12. 37	Sloten, Friesland, HOLLAND (52° 22′ N., 4° 49′ E.).	5. 11. 38
2. 1. 37	Shot, Douces, Maine-et-Loire, FRANCE (47° 25′ N., 0° 30′ W.).	— 11. 37
	Recovered December	
26. 1. 37	Shot near Gouda, HOLLAND (54° 1′ N., 4° 42′ E.).	7. 12. 38
17. 11. 37	Leeuwarden, HOLLAND (53° 13′ N., 5° 48′ E.).	2. 12. 38
37/38	St Jean-le-Thomas, Manche, FRANCE (48° 44′ N., 1° 31′ W.).	30. 12. 38
	Recovered January	
21. 11. 37	Shot, Reeuwyksche Plassen, Gouda, HOLLAND (54° 1′ N., 4° 42′ E.).	16. 1. 38
12. 1. 37	Shot, Mesquer, Guérande, Loire Inférieure, FRANCE (47° 20′ N., 2° 25′ W.).	5. 1. 38
21. 1. 37	Le Pellerin, Loire Inférieure, FRANCE (47° 25′ N., 1° 40′ W.).	29. 1. 38
	Recovered February	
23. 12. 36	Shot, Merckem, near Dixmude, BELGIUM (50° 57′ N., 2° 52′ E.).	25. 2. 38
21. 11. 36	Shot on the River Ticino, Milan, ITALY (45° 27′ N., 9° 10′ E.).	13. 2. 38
	Recovered March	
14. 12. 36	Shot, Ryckevorsel, 28 km. N.E. of Antwerp, BELGIUM (51° 13′ N., 4° 25′ E.).	28. 3. 37
9. 1. 38	Dixmude, BELGIUM (51° 2′ N., 2° 53′ E.).	12. 3. 38
12. 11. 35	Shot, Valle Sacche, Codevigo, Padua, ITALY (45° 23′ N., 11° 54′ E.).	8. 3. 37
	Recovered April	
2. 11. 36	Shot, Lissing Noss, near Leningrad, RUSSIA (60° N., 30° E.).	29. 4. 37
7. 11. 36	Gulf of Finland, near Leningrad, RUSSIA (60° N., 30° E.).	29. 4. 37
16. 12. 37	Arjepluog, SWEDEN (66° 5′ N., 18° 30′ E.).	28. 4. 38
27. 11. 36	Shot, Bulli, LATVIA (57° N., 23° 58′ E.).	23. 4. 38

WIGEON

Recovered May

Ringed		
1. 1. 37	Shot on Vashka River, RUSSIA (64° 30′ N., 40° 9′ E.).	11. 5. 37

Recovered September

22. 2. 35	100 miles N.W. of Kazan, RUSSIA (57° 50′ N., 47° E.).	3. 9. 35
1. 1. 36	Shot, Axai Rostof (Don), RUSSIA (47° 15′ N., 39° 52′ E.).	9. 9. 36
27. 12. 35	Shot, Bulli, LATVIA (57° N., 23° 58′ E.).	23. 9. 38

Recovered October

29. 11. 36	Shot, Nordstrand, North Frisian Islands, GERMANY (54° 30′ N., 8° 50′ E.).	26. 10. 37

Recovered December

10. 11. 38	Shot, St Jacobie Parochie, Friesland, HOLLAND (53° 17′ N., 5° 36′ E.).	18. 12. 38

Recovered January

31. 10. 38	Setubal, PORTUGAL (38° 31′ N., 8° 53′ W.).	12. 1. 39

Recovered April

2. 2. 36	Itlar (Northern Railway) on river Nerll, 100 miles N.E. of Moscow, RUSSIA (56° 50′ N., 39° 15′ E.).	29. 4. 36

PINTAIL

Recovered May

26. 11. 36	Delta of Pechora River, Golodnaya Bay, RUSSIA (67° 45′ N., 52° 30′ E.).	6. 5. 37

Recovered August

5. 11. 37	Vikingavatn, North ICELAND (65° 15′ N., 20° W.).	10. 8. 38

Recovered April

16. 12. 37	Killed by Peregrine, Oostzaan, HOLLAND (52° 25′ N., 4° 50′ E.).	9. 4. 38

SHOVELER

Recovered August

3. 11. 35	Shot, Rügenwalde, Kreis Schlawe, Pomerania, GERMANY (54° 26′ N., 16° 25′ E.).	30. 8. 36
11. 12. 36	Shot, Ijsselmeer, Kampen, HOLLAND (52° 34′ N., 5° 55′ E.).	28. 8. 37
26. 1. 38	Shot, Culemborg, Gelderland, HOLLAND (51° 57′ N., 5° 13′ E.).	25. 8. 38

Recovered October

4. 12. 36	Shot near Dordrecht, HOLLAND (51° 48′ N., 4° 39′ E.).	6. 10. 37

TUFTED DUCK

Recovered June

31. 1. 38	Pechora Delta, Makarovo, RUSSIA (67° 42′ N., 52° 58′ E.).	7. 6. 38

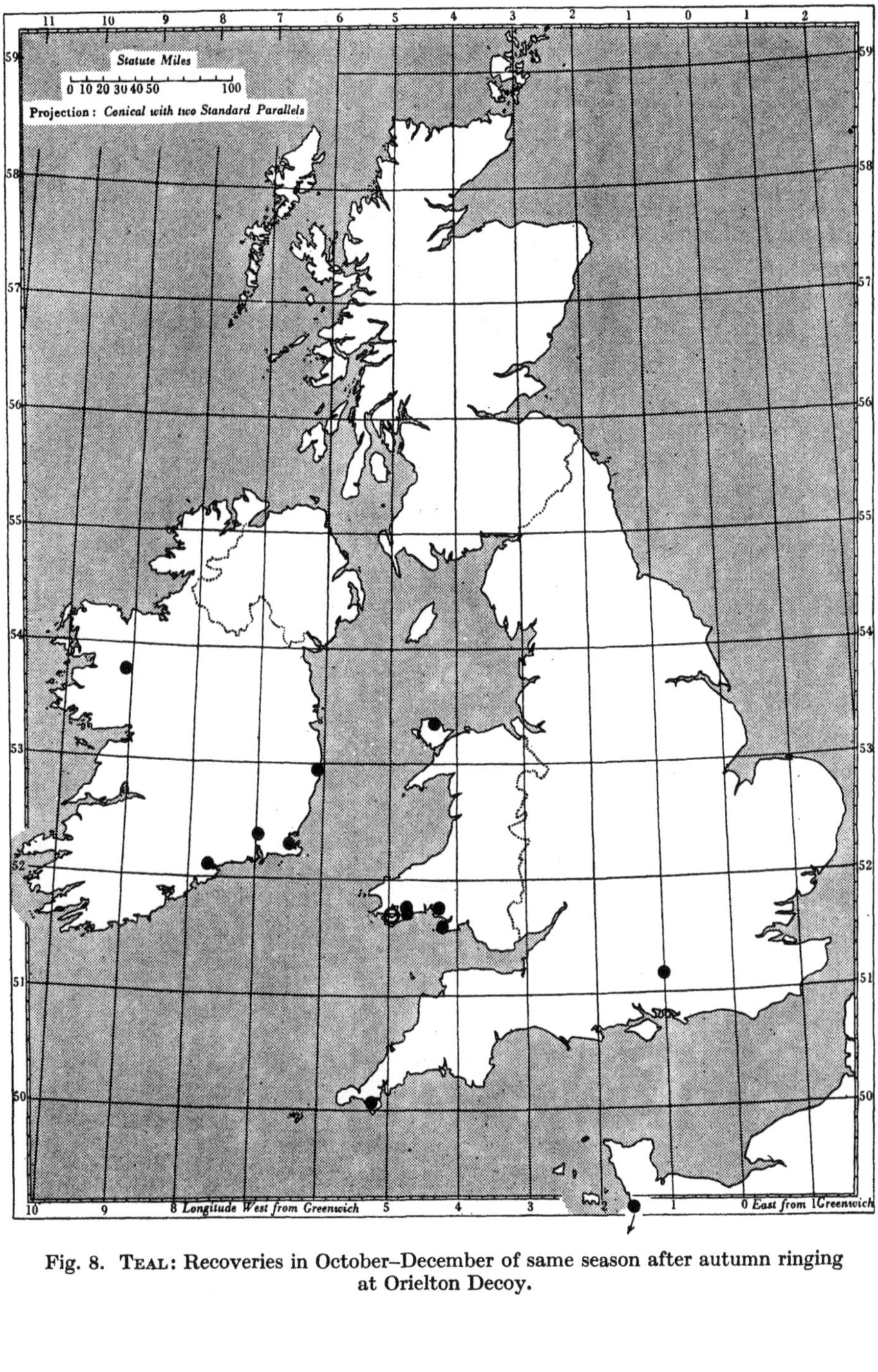

Fig. 8. TEAL: Recoveries in October–December of same season after autumn ringing at Orielton Decoy.

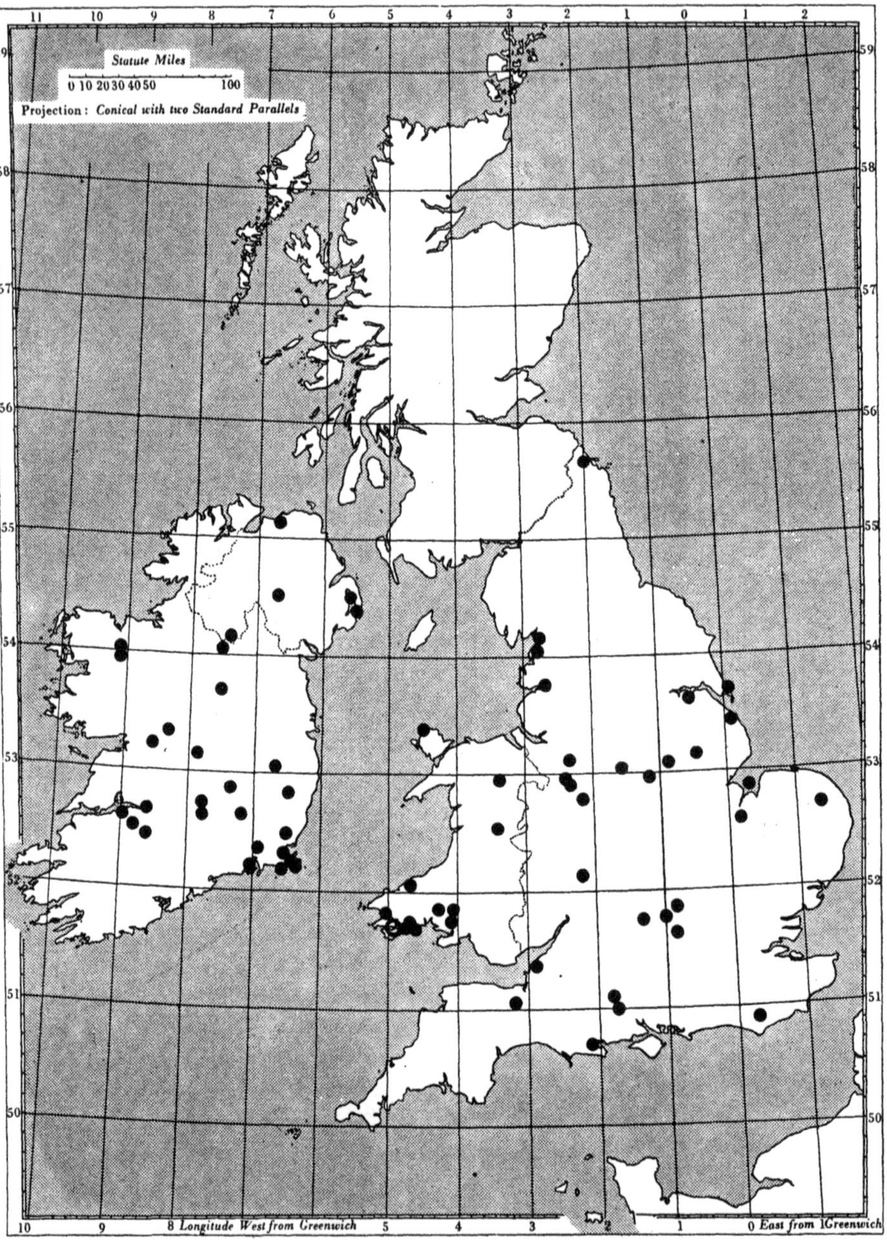

Fig. 9. TEAL: Recoveries in January–February of same season after autumn or winter ringing at Orielton Decoy.

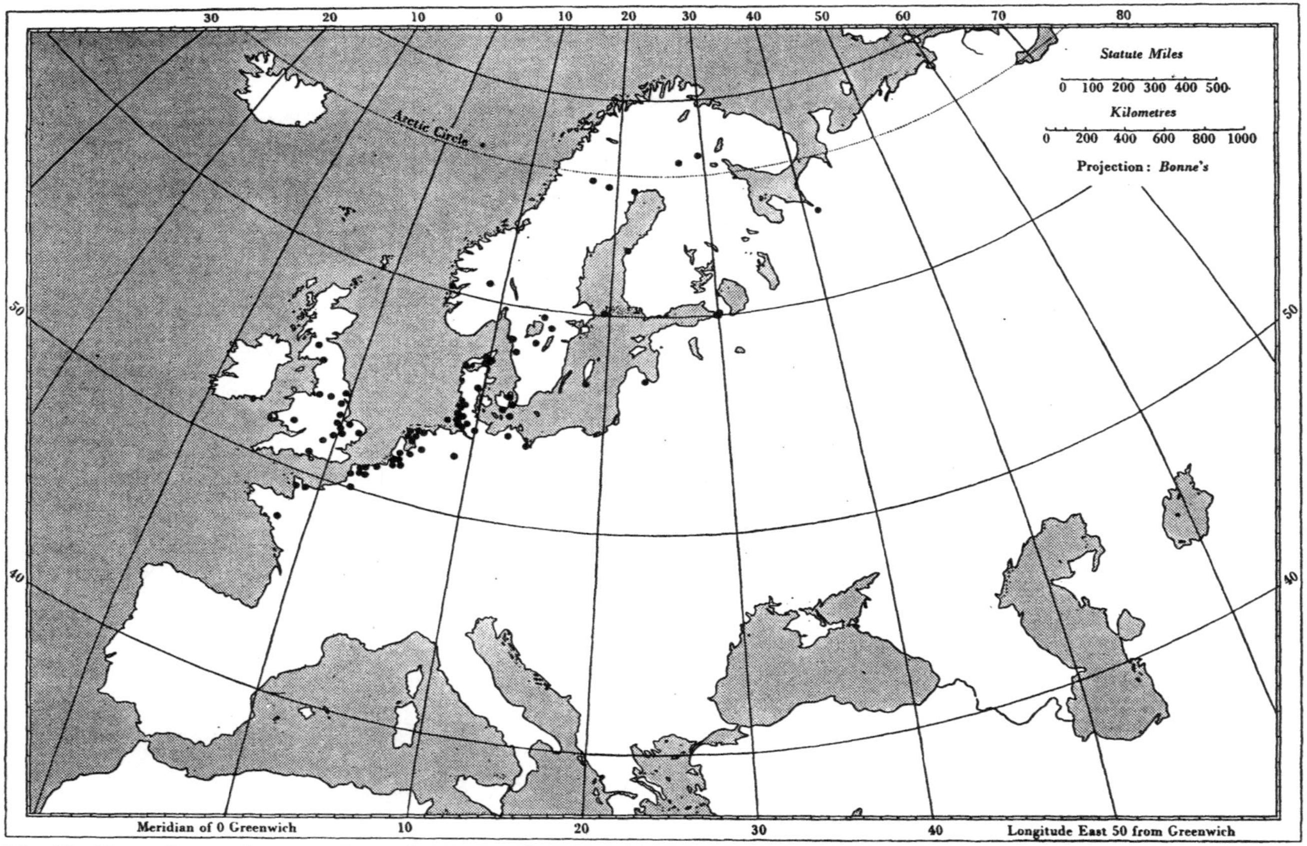

Fig. 10. TEAL: Recoveries on spring and autumn migration of birds ringed at Orielton Decoy. (A dozen recoveries in Holland and a few in the Frisian Islands are not separately shown.)

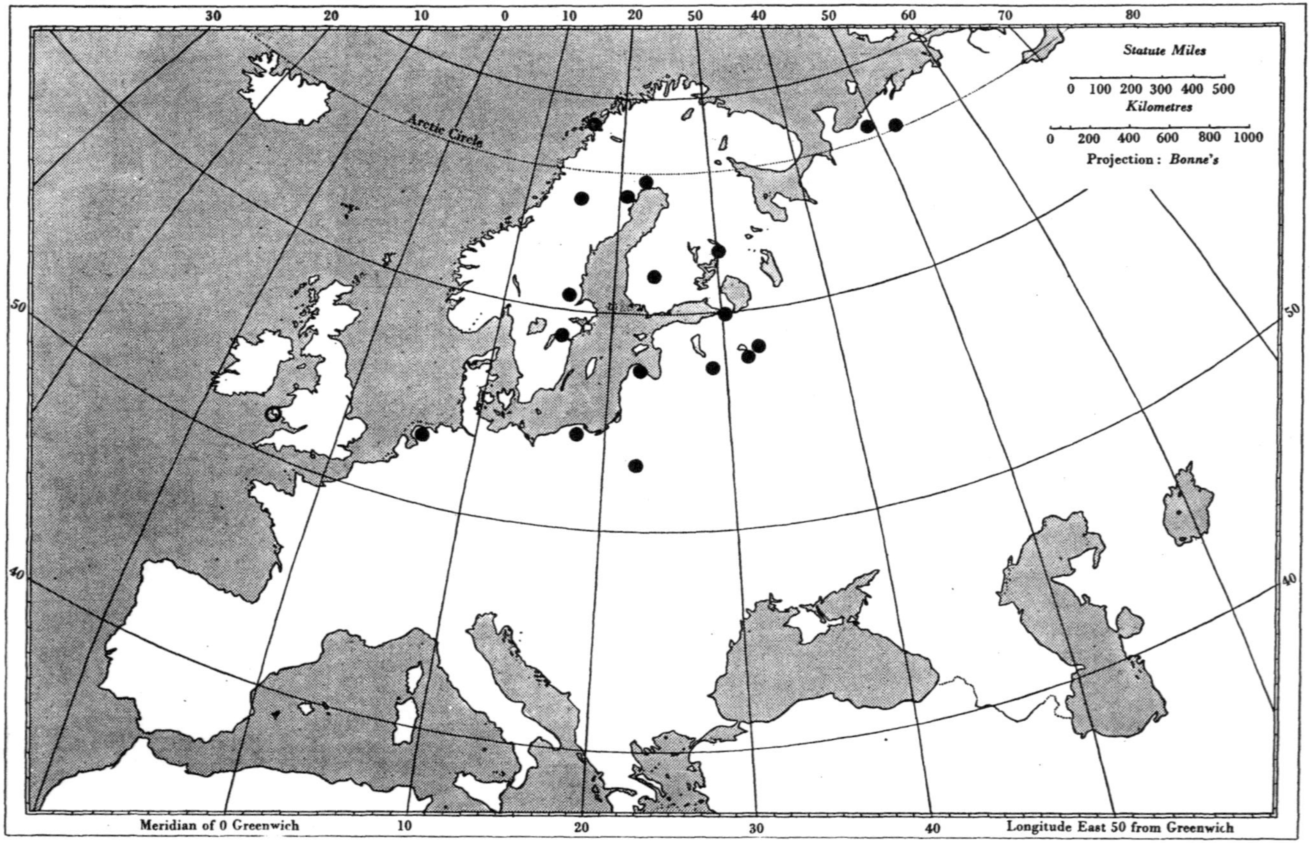

Fig. 11. TEAL: Recoveries in summer (May–August) of birds ringed at Orielton Decoy.

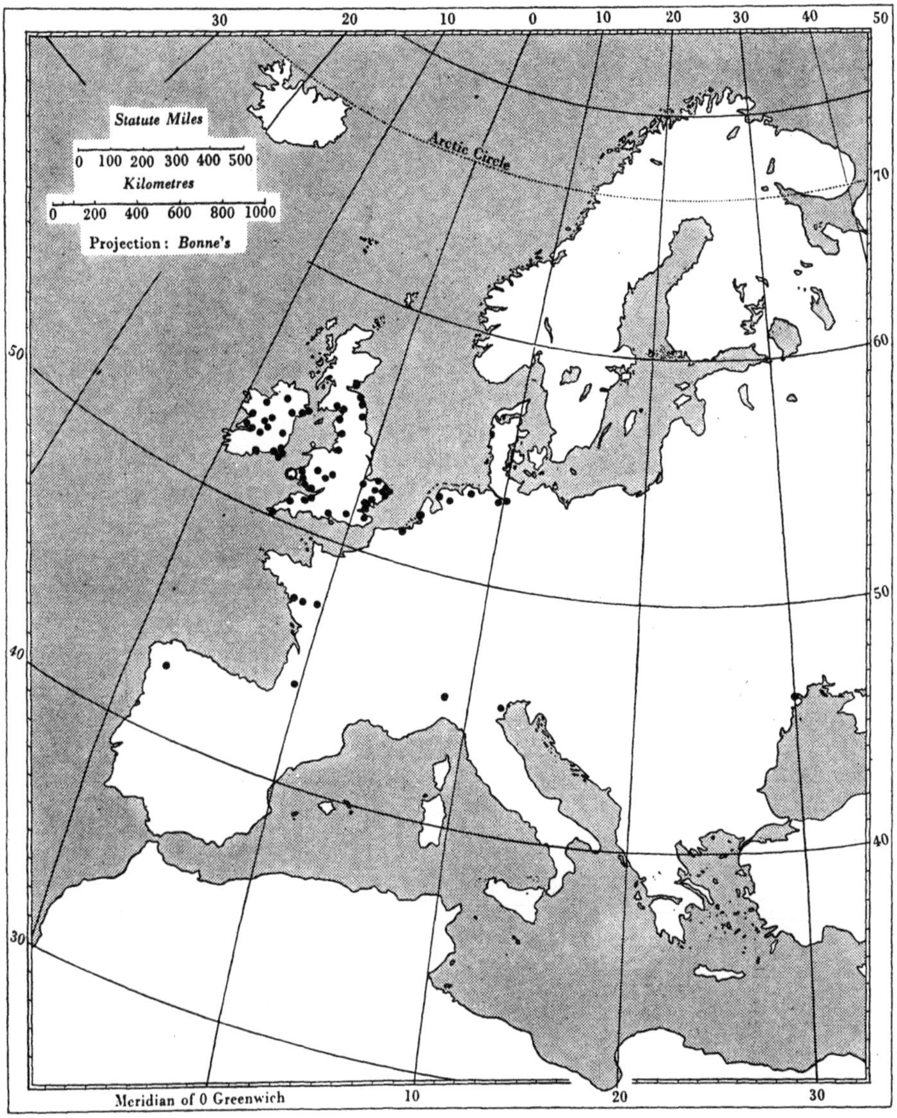

Fig. 12. TEAL: Recoveries in second and subsequent winters (some probably still on migration) of birds ringed at Orielton Decoy. (102 birds recovered at or near the decoy are not shown.)

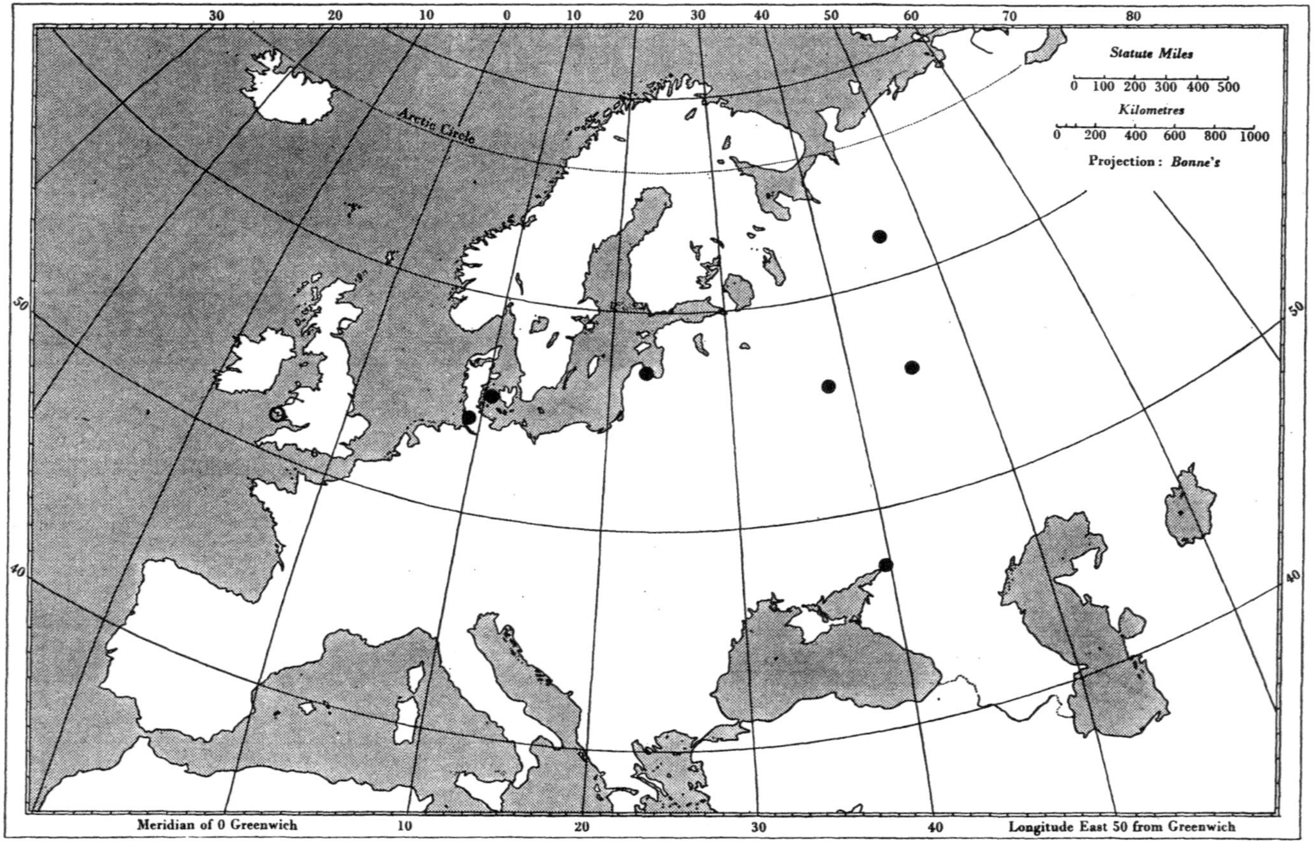

Fig. 13. WIGEON: Recoveries in summer or early autumn of birds ringed at Orielton Decoy.

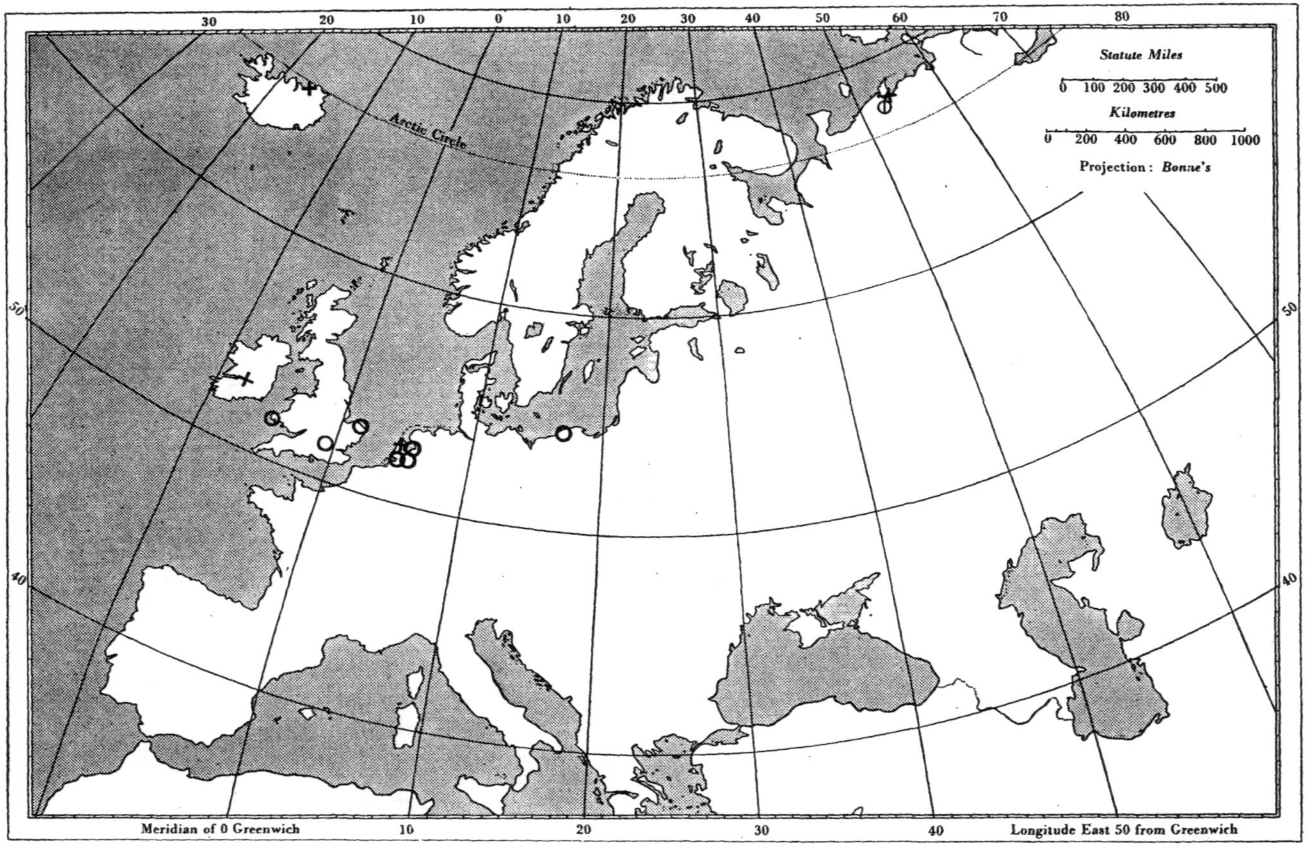

Fig. 14. Pintail, Shoveler and Tufted Duck: Recoveries in summer or autumn of birds ringed at Orielton Decoy.
+ Pintail ○ Shoveler ⏀ Tufted Duck

VII

RESULTS OF RINGING DUCK: GENERAL SURVEY OF DATA FROM ALL SOURCES*

With special reference to the British Isles

By A. LANDSBOROUGH THOMSON

INTRODUCTORY

Ringing Schemes

Duck of several species have been ringed in large numbers since the early days of scientific bird-marking in the first decade of this century. Much interesting information has already been obtained in this way, and the value of the method has been amply demonstrated.

In this country the chief ringing scheme has been that instituted in 1909 by Mr H. F. Witherby in association with his magazine, *British Birds*. The management of this scheme has lately been transferred to a special committee of the British Trust for Ornithology, with headquarters at the British Museum (Natural History). The rings used were formerly inscribed "Witherby High Holborn London", but now bear the address "British Museum Nat. Hist. London"—in addition, of course, to a number for individual identification. Duck have always figured prominently among the birds ringed under this scheme.

Many duck were also marked under the Aberdeen University scheme, which was initiated in 1909 by the present writer but came to an end in the war of 1914–18: the rings were inscribed "Aberdeen University". In recent years duck have been ringed in Pembrokeshire by Mr C. W. Mackworth Praed and Mr H. A. Gilbert with rings inscribed "Orielton Decoy, Pembroke, Britain", but these ringers—like others who are marking duck for the Wildfowl Committee—are now using British Trust rings.

The earliest important results of ringing duck were those obtained by the work of the late H. C. C. Mortensen in Denmark. His reports on the Teal and the Pintail, published in 1908 and 1914 respectively,

* The survey was completed in the middle of 1939, and data becoming available since then are not included.

are the classics of the subject. Since then duck have been marked in large numbers in several other European countries, notably in Iceland, Holland, Germany, Sweden, Finland and Russia. There has also been some ringing of duck in India and in Japan.

In North America, duck-ringing has been practised on a great scale in recent years under the auspices of the U.S. Biological Survey. The results obtained in that continent, however, do not come within the scope of the present review.

Published Sources

The data to be noticed here have all been published, but it will not be practicable to cite all the separate references. These consist of the periodical reports or record lists of the various ringing schemes in the different countries: in respect of the main scheme in this country these are all to be found in *British Birds*. The results obtained in the work done for the Wildfowl Committee are given in detail in chapter VI (pp. 72–76).

Certain collections of data have been particularly useful for the purposes of this survey, however, and must be specially mentioned. For records showing movement between the British Isles and countries abroad, the series of papers by Witherby & Leach (1931 *et seq.*) is invaluable. For records showing movement on the continent, the atlas of Schüz & Weigold (1931), with the subsequent addenda of Schüz (1931 *et seq.*), has been most useful.

Mention may also be made of publications which have dealt specially with duck. There are the two papers by Mortensen (1908, 1914), already mentioned, dealing with Danish results for the Teal and the Pintail; and one by Skovgaard (1939) on recoveries of Icelandic birds in Scotland and elsewhere. There are the reports by Wuczeticz & Tugarinov (1937*a*, 1937*b*) on the Russian results for the Mallard and the Pintail. The earlier British results were summarised by the present writer (1923, 1926). The full references are given at the end of this chapter (p. 105).

Results of Ringing in the Breeding Area

When duck are ringed in the breeding season, usually as young birds, the results give information as to the migratory status of the native population of the species—whether the birds leave the area for the winter, and where they go. Thus, the Mallard native to the British Isles are shown to be more or less stationary; and those native to, say, Finland to be largely migratory, visiting certain other countries in winter. In this country only the Mallard has as yet been ringed in the breeding season on anything approaching an adequate scale, although the Teal and the Wigeon have also been ringed in some numbers.

From the point of view of countries which the birds may visit in winter or on migration, marking in the native area is satisfactory because of the definite nature of the records. There can be no doubt—at least in the case of a record in the first autumn or winter—as to the origin of the particular bird and as to the population to which it properly belongs.

Nevertheless, these records in the bulk may produce a misleading impression at first sight, owing to the unnatural predominance of records relating to birds belonging to breeding areas where marking happens to have been practised on a large scale. Thus, records of birds native to Iceland have in recent years formed a large proportion of the winter recoveries of foreign ducks in the British Isles. These records are of much interest in themselves, but a statistical comparison with records of birds from other sources would be meaningless.

Anomalous Behaviour of Individuals

A difficulty in interpreting the ringing results for duck arises from the fact that there is a good deal of anomalous behaviour on the part of a minority of individual birds—more than is usual among most other migratory species for which data are available. An individual migrant may behave differently in successive autumns; or it may return in spring to a breeding area other than that from which it came, and thus become attached to a different population.

More surprising than such aberrant return, on the part of a migrant, is spring migration on the part of an individual belonging to a stationary population. In that case there is a spring journey by a bird which had presumably remained in its native area throughout the winter, and this bird will thus become an adopted member of another—and probably migratory—population. This irregularity has been shown by ringing results to occur with some frequency among ducks of several species. It has been termed "abmigration", in contradistinction to normal emigration, by the present writer (1923, 1931). It particularly affects the British Isles, where the native birds of species such as the Mallard are practically stationary as a rule, but in some cases accompany winter visitors returning in spring to, for example, Sweden. The accompanying diagram illustrates the definition.

Mention is made below, as regards the Mallard, of an experiment in artificially removing birds from one breeding area to another.

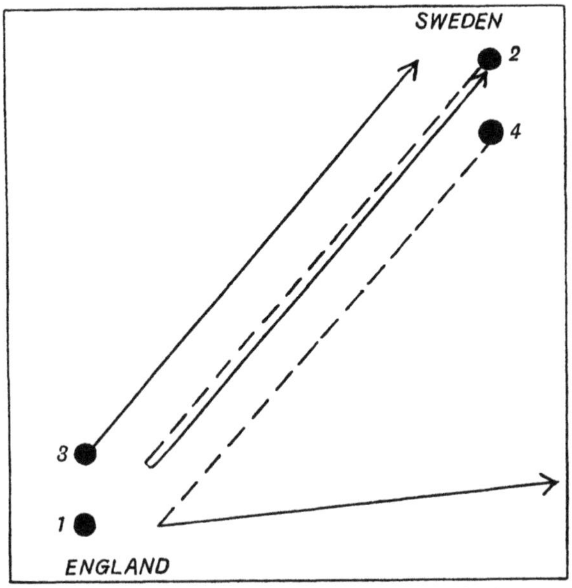

● Birthplace. – – – – Autumn movement. ——— Spring movement.
1. Normal stationary habit. 2. Normal migration. 3. "Abmigration."
4. Migration with aberrant return.

Fig. 15. Diagram to illustrate "abmigration" (*after Thomson*, 1931).

Results of Ringing in Winter Quarters

Ringing of birds caught and released in winter, at decoys or elsewhere, may yield information regarding the origin of visitors to the area at that season—that is, if they are afterwards recorded elsewhere in summer. Provided that the records are sufficiently numerous, the effect of possible "abmigration" may be excluded.

On the other hand, some of the birds so marked may be stationary natives, and records from the same area in subsequent winters are, therefore, of doubtful significance. Records from other areas in subsequent winters may also be difficult to interpret, especially in view of the possibility of "abmigration" having occurred in the interval.

Results of Ringing on Migration

Many duck have been ringed at some midway point on their autumn migration, for instance those caught in Danish decoys or those captured in southern Russia during wing-moult. The winter recoveries show the further course of the migration, and later the summer recoveries show the native areas from which the migrants may be presumed to have come. The results plotted on a map, if sufficiently numerous, give the impression of two fans spreading out from the locality of marking, but of course the apparent concentration of migrants at that point is purely an artificial result of its selection as a centre of ringing activity.

A difficulty of interpretation here arises from the fact that individuals performing quite different migrations may be included. The results give an interesting general picture; but it is not possible to state the native area of any particular bird reeovered in winter, and *vice versa*. Further, birds recovered in winters other than the first may not necessarily have passed through, or even near, the original locality of marking: it is in fact certain that a particular individual may sometimes perform quite a different migration in another year.

MALLARD (*Anas platyrhyncha*)

Status of Native Birds

Mallard have been ringed in considerable numbers as young birds in Great Britain, and the percentage of recoveries has been good. As a result, it has now for many years been definitely known that our native birds are mainly stationary, the great majority of recoveries being from the respective places of marking. In some cases journeys within the British Isles are shown, including crossings from Scotland to Ireland.

It may be mentioned that much the same appears to be true of birds native to Holland: it is largely true also of birds native to Denmark, although some leave there for the winter. Further north the species becomes more definitely migratory, but in many parts some birds remain throughout the winter. In Sweden, for instance, some do so wherever water remains unfrozen: in recent years the number of Mallard wintering in harbours and in city waters has increased, and in these circumstances the birds become relatively tame.

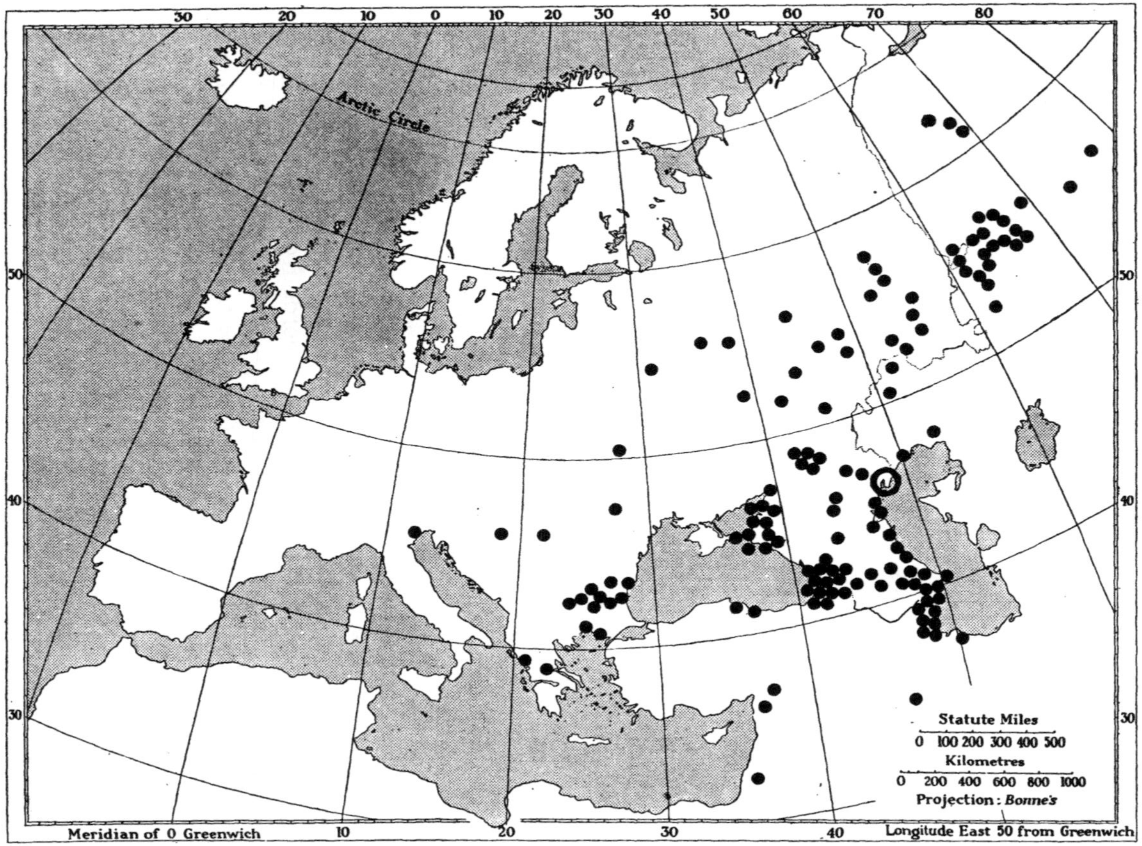

Fig. 16. Mallard from the Volga delta: Recovery localities, at all seasons, of birds ringed during wing-moult in late summer in the Astrakhan Reserve, Volga delta, S.E. Russia (*after Wuczeticz & Tugarinov; adapted*).

Abmigration

Although our native Mallard are mainly stationary in the sense that they do not go further south for the winter, a few of them leave us in the spring to find a new summer area to the north or east. Thus, hand-reared ducklings marked in Aberdeenshire in 1910 were reported from Denmark and Sweden respectively in the following summer, in the former case as a female with a brood of young; and since then there have been other records, both for hand-reared and wholly wild birds, to show that such anomalous behaviour is not uncommon. It is probably due to association—possibly early pairing—between native individuals and winter visitors.

The fact that Mallard of our native stock, although normally stationary, have the capacity to migrate has also been shown by the interesting experiment described by Välikangas (1933). Eighty-six ducklings hatched in Finland from English eggs were ringed, and were found to behave like native Finnish birds—in the fact of migrating, in the direction of doing so, and in return in the following year. Here again association with other birds, those properly belonging to the region, may provide all the explanation that is required.

Origin of Winter Visitors

Numbers of Mallard have also been caught for ringing in winter, particularly at Leswalt in Wigtownshire and at Hickling in Norfolk, and more recently in Pembrokeshire. A good many of these have been recovered in subsequent summers or autumns in Sweden and Finland, and there have been similar records from Norway, Denmark, Holland, northern Germany, northern Poland, Latvia, Esthonia, and the Faroes. Apart from the possibility of abmigration or other irregularity in occasional instances, such birds may be presumed to have been winter visitors from the countries to which they afterwards returned.

This evidence is supplemented by records of birds marked abroad in the breeding season and recovered here in winter. In addition to countries mentioned above, western Russia and Iceland have in this way been shown to provide some of our winter visitors.

There are also a few records of birds marked here in winter and recovered elsewhere in subsequent winters, but these have either been in the breeding area—for example, a bird lingering in Swedish Lapland as late as November—or no further away than northern

France. Records from this country in subsequent winters cannot be interpreted, as the birds may have been either natives or returned winter visitors.

Other Records

The ringing of Mallard in the breeding season in various parts of Europe has provided evidence of migration from the northern countries in a general south-westerly direction. In addition to migration to the British Isles, already mentioned, there are records of purely continental journeys—for instance, from Sweden to Denmark, north-western Germany and Holland; and from various countries to France. In Germany, many Mallard have also been marked in winter at a duck decoy at Celle, Hannover: the recovery localities lie mostly to the north-east and east, as far as northern Russia. Birds marked in winter in Holland, Switzerland and Hungary have also been recovered in Russia.

With the results for western Europe, it is interesting to compare those published by Wuczeticz & Tugarinov (1937*a*) for Mallard ringed in July and August as moulting birds in the Volga delta, at the northern end of the Caspian Sea, a series of 161 recovery records. The breeding area from which these birds are drawn, as indicated by records in subsequent summers, extends from the lower Volga northwards to 62° N. lat. on the Ob, the lateral extremes being in the Ukraine and in Siberia, 77° E. long. Of the winter records showing further migration, most are from the southern Caspian, the Caucasus region, and the Black Sea: there are also records from Asia Minor, Syria and Palestine. The remainder show a more extended westerly movement—to Bulgaria, Greece, Rumania, and Jugo-Slavia, with one from Italy (near Venice).

There are a few records of Mallard marked in winter in India and in Japan being recovered in summer in different parts of Siberia. In North America the Mallard performs long migrations, and numerous ringing results have been obtained by the U.S. Biological Survey.

TEAL (*Anas crecca*)

Status of Native Birds

Marking of young Teal in the British Isles has so far yielded only a few records. Some show movement within the country, including journeys from Scotland to Ireland and one from Scotland to the south of England. One record shows emigration—a bird ringed when young in Durham and recovered in western France in its first winter.

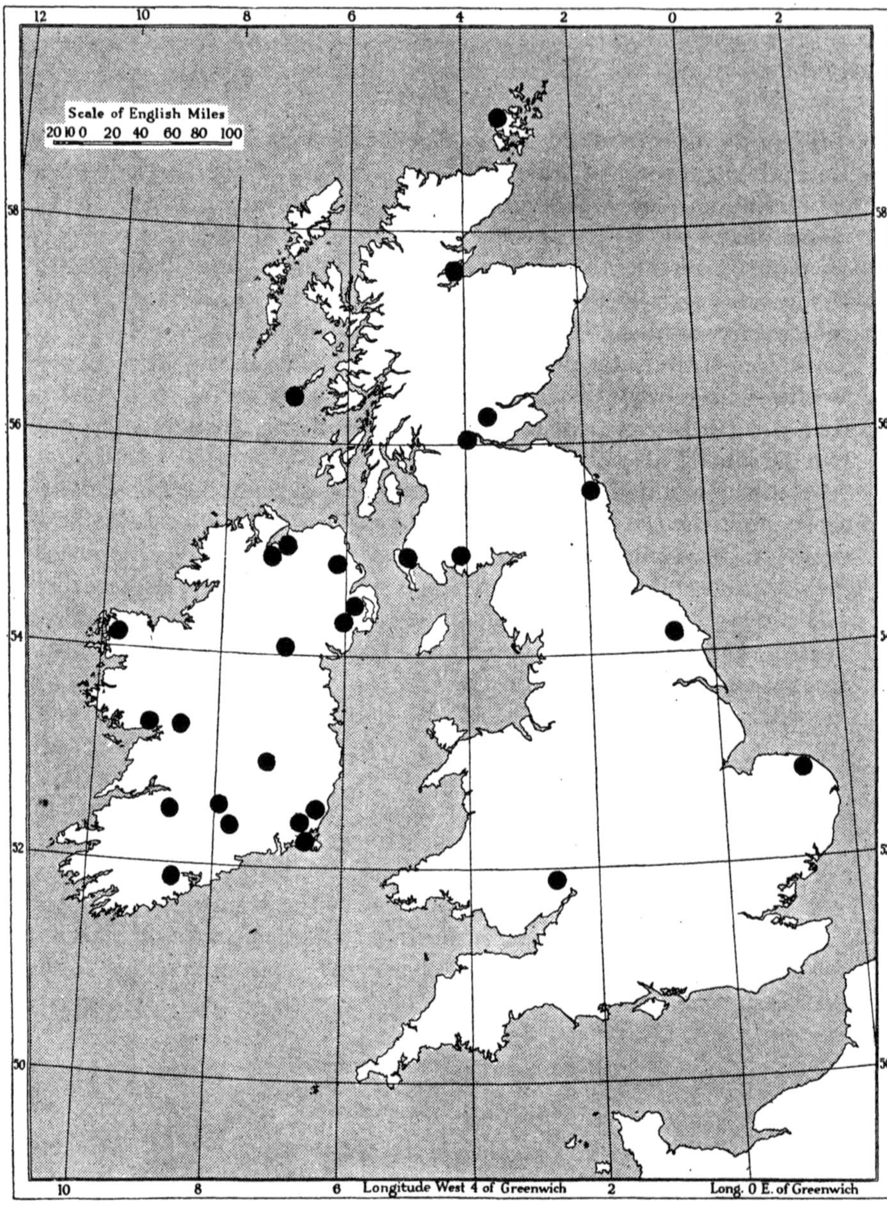

Fig. 17. TEAL FROM ICELAND: Recovery localities (approximate) in the British Isles, in winter, of birds ringed in Iceland in the breeding season, mainly as young.

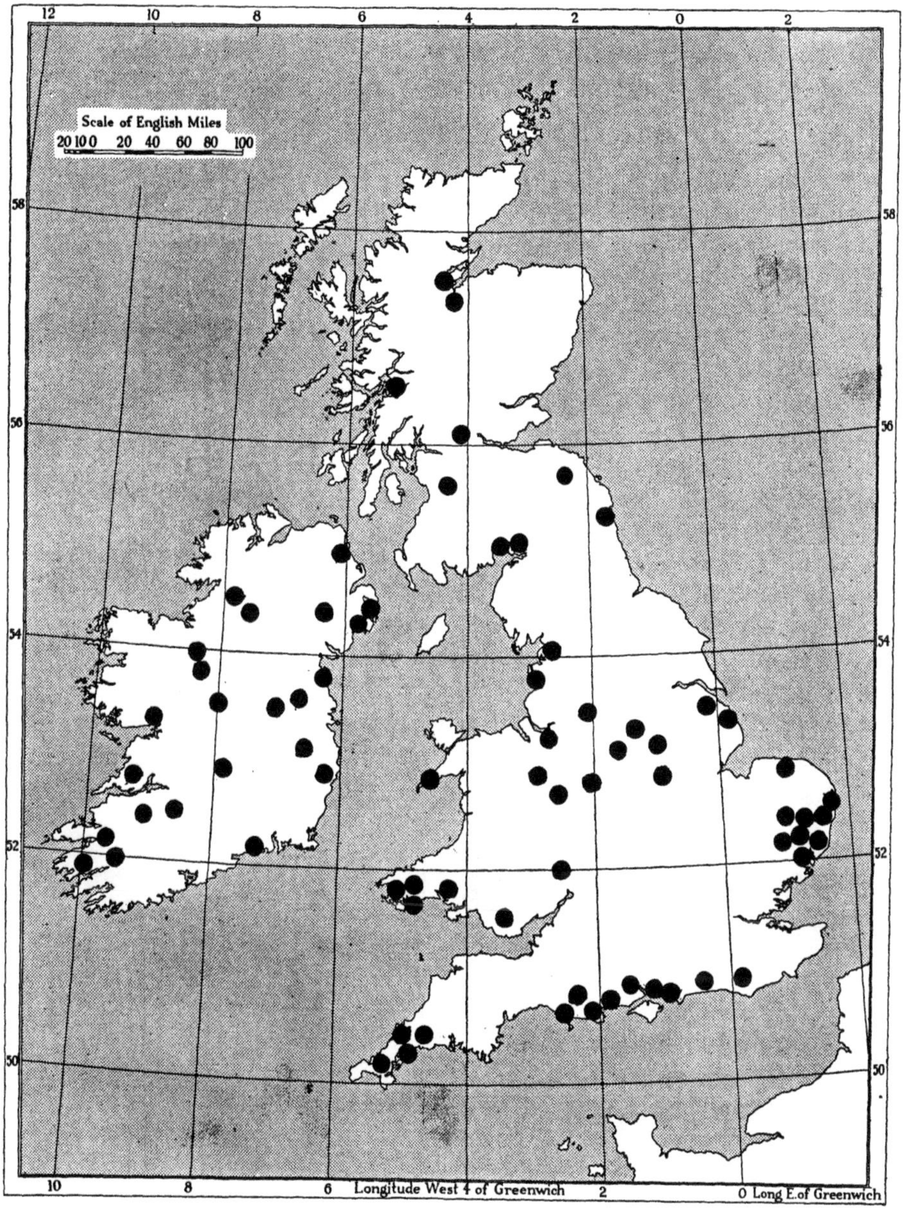

Fig. 18. TEAL FROM A DANISH DECOY: Recovery localities (approximate) in the British Isles, in winter, of birds ringed in autumn at a decoy on the island of Fanö, western Denmark.

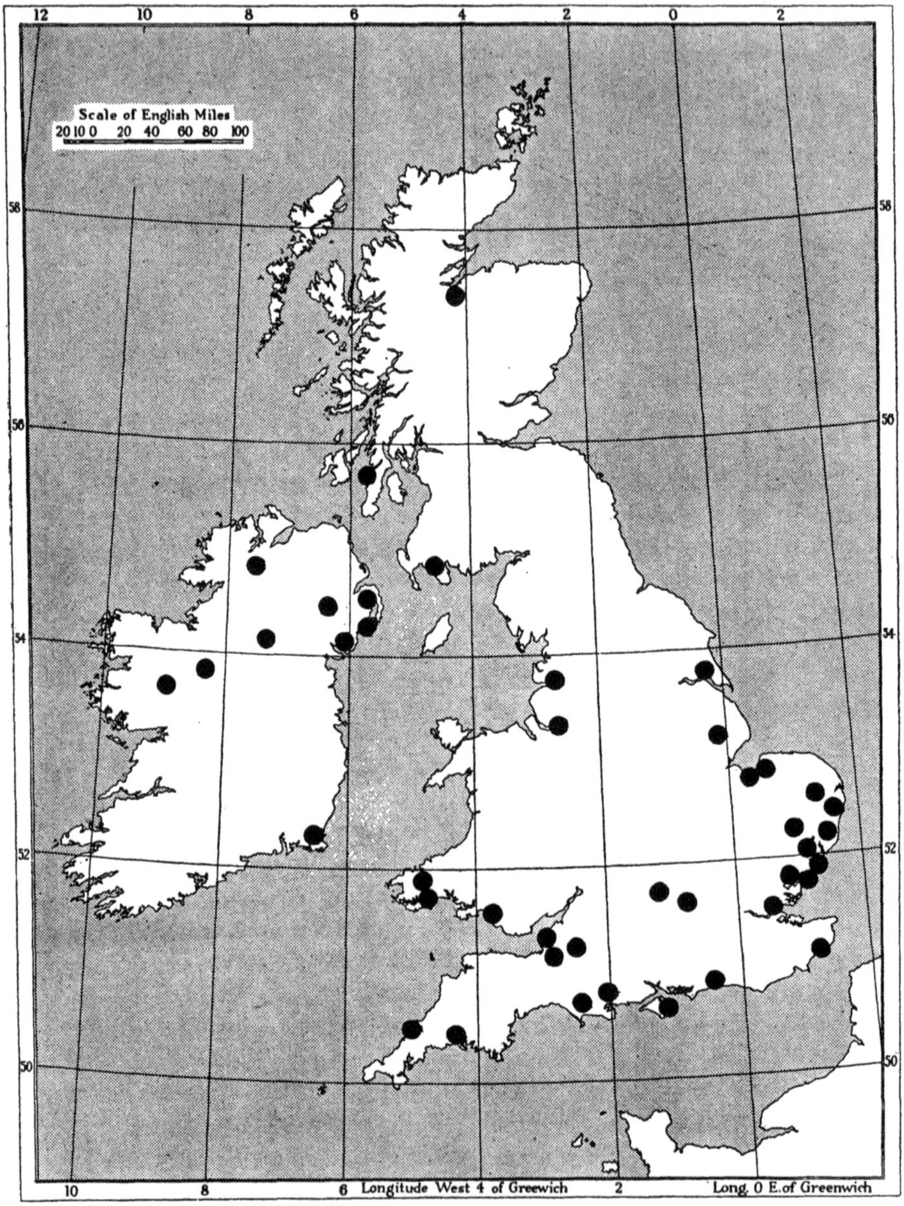

Fig. 19. TEAL FROM DUTCH DECOYS: Recovery localities (approximate) in the British Isles, in winter, of birds ringed in autumn or winter at decoys in Holland.

Abmigration

Records showing this phenomenon are particularly numerous in the case of the Teal, although it must be said that most of the birds were hand-reared and therefore conceivably more likely to behave in an abnormal manner. These records are of birds ringed as young in this country and recovered abroad in subsequent summers—mostly in Sweden, but also in Russia, Norway, Denmark and northern Germany.

Origin of Winter Visitors

Teal ringed in Great Britain in winter have been recovered in summer or autumn in Norway, Sweden, Finland, Latvia, Russia, Poland, Germany, Denmark, Holland and France. Most of the records are included in the Orielton list in chapter VI, which should be consulted for details.

Teal ringed in the breeding season in Iceland, Sweden, Finland, Russia and Latvia have been recovered in the British Isles in winter; and also birds marked at decoys in autumn in Denmark and Holland. The most numerous records are those relating to birds from Iceland and from Denmark, marking having been particularly active in those countries. In the case of Icelandic birds, the recoveries in Ireland outnumber those in Great Britain. Birds marked in Denmark on autumn passage have been recorded in greatest numbers in the west of England and Wales, but also largely in Ireland, as well as to a less extent in the eastern parts of England and in Scotland.

Records of birds ringed in winter and recovered in subsequent winters are always difficult to interpret, as the breeding origin is unknown. Where the record is from the same place, the bird may be either a sedentary native or a returned winter visitor; where the record is from a distance, the bird was probably a winter visitor in the first instance (unless it was subsequently an abmigrant). There are several cases of Teal ringed in Great Britain in winter being recovered in Holland, Belgium or France in subsequent winters. In cases where recovery was in northern Spain (2), in northern Italy (2), and in Rumania (1), it may be supposed that a different migration was performed than in the season of marking.

Other Records

Recoveries of Teal marked in the breeding season in various northern countries of Europe show movement not only to the British Isles but in a general south-westerly or southerly direction on the Continent—to France, Spain and Italy. The same is true of birds ringed in autumn at the decoy on Fanö, on the west coast of Denmark, by Mortensen many years ago, and more recently by Skovgaard. Icelandic birds have been recovered in Sweden, Finland, Russia, Belgium, France and Portugal.

Over 300 Teal were ringed by Mr H. H. Buisman in the autumn of 1937, at a decoy at Piaam, in Friesland, Holland, and forty were recovered before the end of the winter. The recoveries were mainly in England (especially the eastern counties) and France, with a few from Ireland, Scotland, Germany and Holland itself. The ringing work was continued until the middle of December, and it is of interest that birds marked as late as that were among those which travelled to England and France. There have since been summer recoveries in Denmark, Sweden and Russia. Further ringing at Piaam in 1938 has yielded additional records, including winter recoveries in Spain and Portugal.

WIGEON (*Anas penelope*)

Status of Native Birds

Wigeon native to the British Isles have not been marked in large numbers, but there is evidence that at least some migrate southwards in winter. A young bird ringed in Sutherland in 1909 was recovered in Holland early in September of the same year; another member of the brood was recorded from Nottinghamshire in the next winter. Two others have travelled from Scotland to Ireland. In the more northerly parts of its range this species is definitely migratory: it is, for instance, almost entirely absent from Sweden in winter.

Abmigration

Either abmigration in the strict sense, or else return—after southward movement—to another breeding area, is shown by two cases of young birds ringed in this country and recovered in subsequent summers in northern Russia: one of these was hand-reared

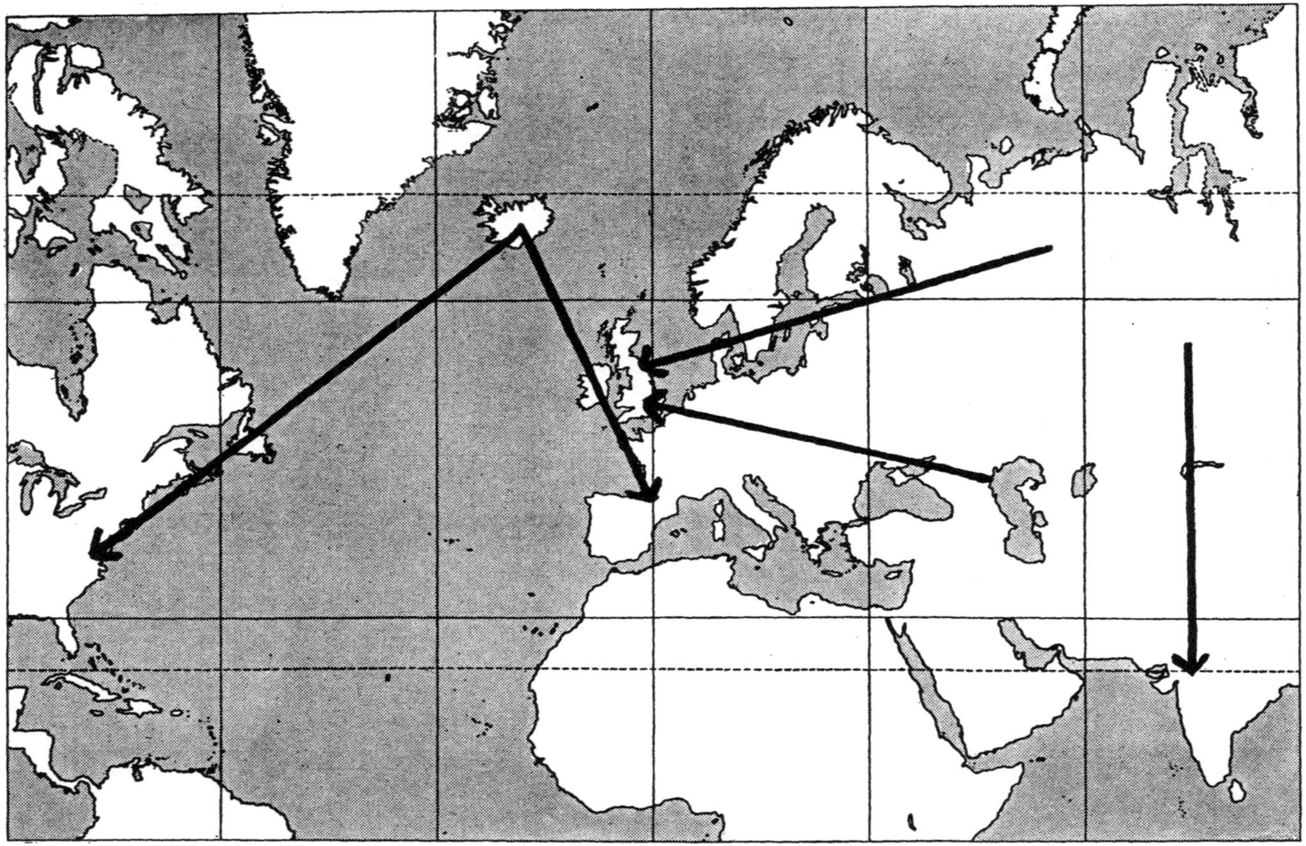

Fig. 20. Some directions of Wigeon migration: A diagrammatic representation of some of the chief directions of autumn migration which have been shown by records of ringed birds.

and the other a wild bird. It is difficult, however, to interpret the case of a young wild bird ringed in Kinross-shire and recovered far to the north, in Iceland, in January of its first winter: another bird ringed in Kinross-shire had travelled northwards to Cromarty by August of the same year.

Origin of Winter Visitors

Birds ringed in winter in Pembrokeshire (see chapter VI, p. 76) and elsewhere in Great Britain have been recovered in the breeding season in Denmark and Russia, and in Germany and Latvia in autumn. A remarkable case is that of a Wigeon ringed in Warwickshire in October 1915 and recovered in Uralsk, south-eastern Russia, in April 1918. Also of special interest are the birds ringed in Pembrokeshire and recovered in Holland and Portugal, respectively, later in the same winter: one ringed in Norfolk was recovered in Italy in the next winter.

There are also a few records of birds ringed on the Continent—in East Prussia and Latvia—in the breeding season and recovered in this country in winter. Others, ringed at decoys in Denmark and Holland, have also been recorded here. Two others, ringed during moult in late summer in the Volga delta, have been recovered in Suffolk and in Co. Kildare, respectively, in December of the same year.

More numerous, although presumably only because marking has been done on a greater scale, are the records of birds ringed in Iceland in the breeding season, mostly as young birds, and recovered in the British Isles between September and February of the first or succeeding winters.

Other Records

Although most of the recoveries of Iceland Wigeon are from the British Isles, some of these birds have been recorded from other parts of Europe—including records from southern Spain, northern Italy, and central Russia—and one from Tunis in North Africa. Still others have reached Newfoundland and the eastern seaboard of North America, and there is one record from Antigua in the West Indies.

There are also a few recovery records in northern France of birds ringed in Holland and Denmark.

Wigeon marked in India in winter have been recovered in Siberia in summer.

PINTAIL (*Anas acuta*)

Status of Native Birds

Such little ringing as has been done in the case of native British birds has yielded only three records, one of which shows movement from Scotland to Ireland in the first winter.

Origin of Winter Visitors

A Pintail ringed in Cumberland early in March was recovered in Schleswig-Holstein in September of the same year. Three ringed in Pembrokeshire have been recovered, respectively, in Holland in spring, and in Iceland and Russia in summer, as shown in the previous chapter.

Several Pintail marked as young in Iceland have been recovered between November and January in the British Isles—the majority in Ireland. Birds ringed at a decoy on Fanö, Denmark, have also been recovered in Scotland and Ireland.

Other Records

The ringing of Pintail at the decoy on Fanö, at the south-western corner of Denmark, has yielded many interesting records in addition to those already mentioned from the British Isles. The series published by Mortensen (1914) resulted from the marking of 320 birds of this species caught and released in the autumns of 1908, 1909 and 1910. Of these no less than sixty-seven, or over 20 per cent, were recovered—two of them twice and most of them at a distance.

The summer recoveries of these Fanö birds show that they were drawn from a breeding area which includes Finland and Sweden, and which extends to northern Russia as far as the White Sea and the Arctic Ocean. The recoveries between October and March show further migration to the British Isles, Holland, France, Spain, Italy and Jugo-Slavia. The distribution of the records is shown on the accompanying map (Fig. 21); a more detailed summary than that given here has been published elsewhere by the present writer (1926).

Birds marked in Iceland in summer have travelled to other countries than the British Isles, there being single records from Canada, Sweden, Germany, Holland, southern Italy and Albania. Birds marked in Sweden have been recovered in France, northern Italy, and Algeria. Other journeys recorded are from eastern Den-

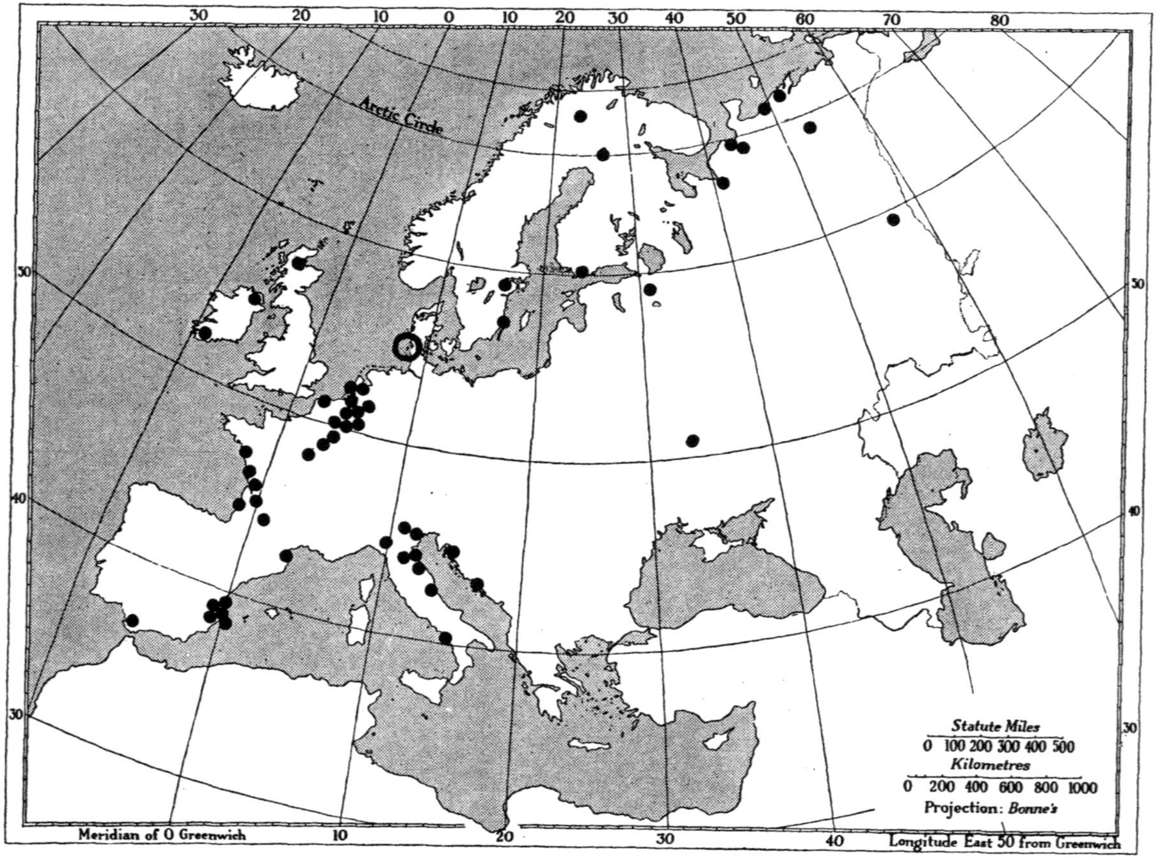

Fig. 21. PINTAIL FROM A DANISH DECOY: Recovery localities, at all seasons, of Pintail ringed in autumn at a decoy on the island of Fanö, western Denmark (*after Mortensen; adapted*).

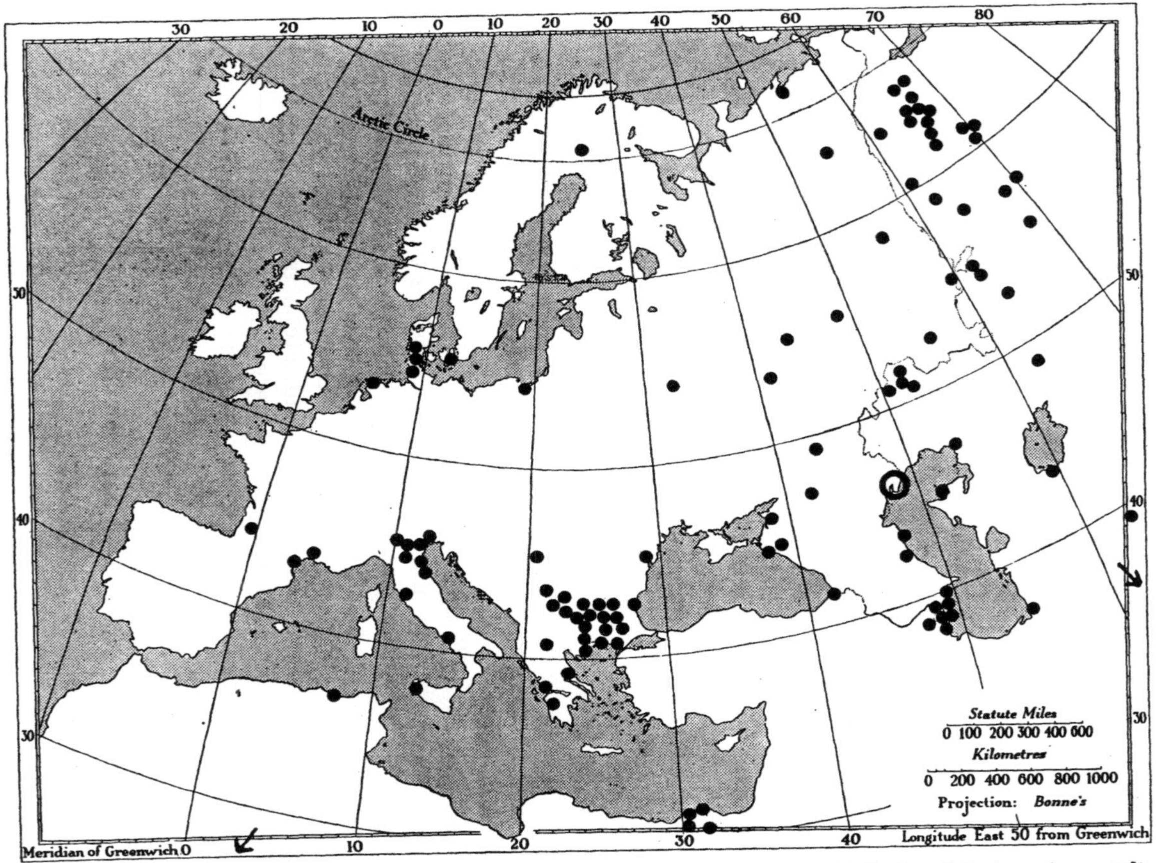

Fig. 22. Pintail from the Volga delta: Recovery localities, at all seasons, of birds ringed during wing-moult in late summer in the Astrakhan Reserve, Volga delta, S.E. Russia (*after Wuczeticz & Tugarinov; adapted*).

mark to Holland and Spain; and from Holland to northern and western France in winter, and to Arctic Norway in summer.

Wuczeticz & Tugarinov (1937*b*) have published a most interesting series of 121 recovery records of Pintail ringed in late July, mostly as adult males in wing-moult, in the delta of the Volga at the head of the Caspian Sea. The species, it is stated, does not usually breed in this area, and the birds have therefore already performed part of their migration before they are ringed—and it is noteworthy how far some of them must have travelled between breeding and the "eclipse" of the summer moult. The native area of the birds is indicated by recoveries in subsequent summers. It lies to the northward, comprising the Ural region of Russia and part of western Siberia; it extends to the Arctic coast and eastwards to the seventieth meridian. Two of the records are from further west, White Russia and Finland respectively.

The records indicating migration routes or winter quarters of these Volga birds fall into two main groups. Firstly, some show comparatively short southward journeys to the Caucasus region and the southern part of the Caspian, and one an eastward journey to the Aral Sea. Secondly, more numerous records show an extensive movement south-westwards and westwards—to southern Russia, Bulgaria, Greece, Rumania, Hungary, Jugo-Slavia, Italy (including Sicily), southern France, Egypt and Algeria. In addition there are two records showing a south-easterly trend, from Turkestan and northern India respectively, and one from French West Africa (on the Niger). There are also a few records from northern Germany, Denmark and Holland: except in the Dutch case, these are all for subsequent seasons and probably indicate a migration different from that performed in the year of marking.

A bird ringed in India in winter has been recovered in Siberia in summer.

OTHER SPECIES

In the case of other species of duck than those treated above, the ringing data are as yet too scanty to be of more than suggestive value. The more interesting records may, however, be briefly noticed. Some of the species have been ringed very little except in Iceland, and the apparent preponderance of that country as a source of our winter visitors must be discounted to some extent for that reason.

Shelduck (*Tadorna tadorna*).

There are several records, both British and German, indicating the existence of considerable intermigration between Great Britain —including its west and south coasts as well as the east—and north-western Germany. The nature of the movement is as yet obscure, and the dates of some of the records make them difficult to interpret.

Gadwall (*Anas strepera*).

A few have been ringed in this country in summer and recovered at the same place in winter. A few ringed in Iceland in summer have been recovered in the British Isles in winter.

Garganey (*Anas querquedula*).

One ringed at a decoy in Holland in August was recovered in Ireland in November of the same year.

Shoveler (*Spatula clypeata*).

One ringed as a duckling in Lancashire has been recovered in Holland in a subsequent winter. Two birds ringed in winter have been recovered in summer in Denmark and northern Germany respectively. Three ringed in winter in Pembrokeshire, as shown in the previous chapter, have been recovered in late summer or autumn in Holland. A native Dutch bird has been recovered in Norfolk, and another in Kirkcudbrightshire (in August of its first year).

Other journeys recorded are from western Russia to southern Germany, Holland and France; from Sweden to France; from Germany to Holland and France, and from Czechoslovakia to northern Italy. Birds marked in south-eastern Russia, at the head of the Caspian Sea, have been recovered in Italy and in Libya.

Pochard (*Aythya ferina*).

A bird ringed in winter in Warwickshire has been recovered in northern Germany, and a bird from Finland has been reported in Buckinghamshire. Other records include several of birds marked in western Siberia in summer and recovered in Italy, and one of a bird ringed in northern India in winter and recovered in Siberia in summer.

Tufted Duck (*Aythya fuligula*).

Several marked as ducklings in this country have been recovered here in winter. Another—a case of "abmigration"—was recovered in northern Finland two summers later.

A bird marked in winter in Pembrokeshire was recorded in the following summer from the Pechora delta, on the north coast of Russia. A bird which was caught in two successive winters on the same reservoir in Surrey was recovered thereafter in spring in southern Finland. Another ringed in winter in Surrey was recovered in northern France in the following summer.

Tufted Duck have been ringed in some numbers in summer in Iceland, and a good many of these have been recovered in winter in the British Isles, especially in Scotland and Ireland. Birds ringed in summer in Finland, in Sweden and in Denmark have also been recovered here in winter. There are single records of Icelandic birds from Norway, Denmark, Holland, France and Italy.

Scaup (*Aythya marila*).

This species has likewise been ringed in considerable numbers in summer in Iceland: a good many of these birds have been recovered in winter in the British Isles—especially in Scotland and Ireland—and also in Holland; and there are records from the Faroes and from Germany. A bird from the Aland Islands, off the Finnish coast, has been recorded from Dalmatia.

Golden-eye (*Bucephala clangula*).

There is a record of a bird marked in western Russia and recovered in the East Frisian Islands.

Long-tailed Duck (*Clangula hyemalis*).

The records for this species are few but interesting. One ringed in Iceland was recovered on the west coast of Greenland a year later. Two ringed in Swedish Lapland have been reported in winter from western Norway and from the Black Sea respectively.

Common Scoter (*Melanitta nigra*).

One ringed in Iceland has been recorded from the Azores—a remarkable journey. Another has travelled from Iceland to Portugal.

Goosander (*Mergus merganser*).

Birds ringed in England in winter have been recovered in Sweden (two) and Finland in summer.

Red-breasted Merganser (*Mergus serrator*).

Birds ringed in summer in Iceland have been recovered in winter in Scotland—also one in Holland.

REFERENCES

MORTENSEN, H. C. C. "Teal (*Anas crecca* L.) in winter." *Vidensk. Meddel. naturh. Forenings Köbnhavn*, p. 127 (1908).

— "Maerkede Spidsaender." *Dansk Ornith. Forenings Tidsskrift*, VIII, 113 (1914).

SCHÜZ, E. "Ring-Wiederfunde ausländischer Stationen. 2." *Vogelzug*, II, 127 (1931 *et seq.*) (and subsequent similar papers in the same journal).

SCHÜZ, E. & WEIGOLD, H. *Atlas des Vogelzuges nach den Beringungsergebnissen bei palaearktischen Vogeln.* (*Abh. Vogelzugsforsch. Helgoland*, No. 3), Berlin (1931).

SKOVGAARD, P. "Iceland birds visiting Scotland." *Scottish Naturalist*, p. 61 (1939).

THOMSON, A. L. "The migrations of some British ducks: results of the marking method." *British Birds*, XVI, 262 (1923).

— *Problems of Bird-Migration*, London and Boston (1926). (Chapter XIII: "The migrations of the Mallard and the Pintail.")

— "On 'abmigration' among the ducks: an anomaly shown by the results of bird-marking." *Proc. VII Internat. Ornith. Congress*, Amsterdam, p. 382 (1931).

VÄLIKANGAS, I. "Finnische Zugvögel aus englischen Vogeleiern." *Vogelzug*, IV, 159 (1933).

WITHERBY, H. F. & LEACH, E. P. "Movements of ringed birds from abroad to the British Islands and from the British Islands abroad." *British Birds*, XXV, 110 (1931 *et seq.*) (and subsequent addenda in the same journal).

WUCZETICZ, W. & TUGARINOV, A. "Seasonal distribution and migration of Ducks on the base of bird ringing in the U.S.S.R. I. The Mallard—*Anas platyrhyncha* L. II. The Pintail—*Dafila acuta* L." Moscow (1937*a*, 1937*b*). (Pamphlets in Russian and English: reviewed and summarised by A. L. T., *British Birds*, XXXII, 20 (1938).)

VIII

CLOSE TIME

By PHYLLIS BARCLAY-SMITH

An adequate "close time" is the fundamental of the conservation of any game bird, and this is recognised by sportsmen in most countries of the world. The question of "close time" for migratory birds is of even greater importance, for there has been a tendency in the past to take the line of "shooting while the shooting is good", and to disregard the interests of neighbouring nations. This attitude is fortunately changing for the wide international view-point in which the well-being of all is considered and the fact of interdependence is acknowledged.

The exact dates of the length of "close time" is a vexed and difficult question, and one which must inevitably depend to a great extent on climatic conditions and geographical position. The main essential in all cases, however, is that birds should be given complete protection from the time they are paired and ready to nest to the time the young are sufficiently strong on the wing to defend themselves. It is of particular importance that migratory birds should be protected while on their journey north to breed. If shooting or trapping is continued too late in the spring, much harm may be done by the destruction of potential parents, and the consequent drain on the stock may do irreparable damage. By commencing the shooting season too early in the summer, when many young birds are not able to fly, masses of flappers are slaughtered; they are too small to be of much food value, they are shot at a time of year when in most countries the weather is so warm as to cause great wastage by decaying flesh, and, most important of all, by killing such large numbers of young birds "new entry" and new blood is lost, to the detriment of the stock as a whole.

In Great Britain there are two sets of laws dealing with the birds of the British Isles—the Game Acts, which originated in the fifteenth century, and the Wild Birds Protection Acts, the first of which was passed in 1880. Wild geese and wild duck are considered by the law to be "wild birds" and, therefore, do not enjoy the privileges accorded to game birds. Under the Wild Birds Protection Acts a "close time" was fixed from March 2nd to July 31st, but County

Councils were given power to alter this period, either by extending it or curtailing it, for the whole or any part of the county. Under this system the "close time" for ducks and geese varied all over Great Britain, and such cases existed that on one side of a river it was legal to shoot duck eleven days longer than on the other.

The Wild Birds (Ducks and Geese) Protection Act, promoted by the British Section, International Committee for Bird Preservation, which came into force on August 1st, 1939, lays down a *minimum* "close time" from February 1st to August 11th, giving power to County Councils to apply for permission to extend, but not, as heretofore, to contract, this period. A uniformity of a minimum "close time" throughout the country has, therefore, at least been achieved. A clause, however, was inserted granting power to County Councils to make application, should they consider fit, to allow shooting on the "parts contiguous to the low-water mark of ordinary tides" up to any date *not later* than February 21st.

This clause was inserted mainly on account of the shooting of Wigeon and Brent Geese, as it was contended that in many districts very little, if any, shooting of these birds is possible till February. Also as Brent Geese and Wigeon do not breed till the end of May, or later, their need for protection during the first half of February is not so great as that of the earlier breeding species. As it is impracticable to attempt to protect one species of duck while allowing the shooting of another species of very similar appearance, it was decided to grant facilities for shooting to continue in the areas occupied by Brent and Wigeon (i.e. maritime) to a later date than in those areas occupied by the earlier nesting duck (i.e. inland).

The Act of 1880 contains a list of birds placed on a Schedule. Under this Act owners and occupiers may not kill or take the birds scheduled even on their own land during the "close time". Wild duck were included in this Schedule, but wild geese were not, so could be legally shot at any time of year on private ground if the owner's consent was given. By the Act of 1939 wild geese (all species) are also placed on the Schedule, and thereby given protection from owners and occupiers of land during the "close time". In order to safeguard the interests of farmers and others who may suffer damage to crops from geese, power is given to County Councils to make application to remove geese from the Schedule in those counties, or parts of counties, where this is found necessary.

The Act includes all species of geese and duck (except Merganser and Goosander), but in the terms of the Act certain species of duck are mentioned and this may lead to confusion. The reason is that in the Act of 1880 certain species of duck were specifically mentioned, and according to Parliamentary custom these must be repeated in any subsequent Act in which they are included, although they are automatically comprised in the words "All species".

The clause of the 1939 Act prohibiting the importation of wild geese and duck from the period February 1st to August 11th is aimed at assisting neighbouring countries in the preservation of their wildfowl during the breeding season, by cutting off the demand from the London markets during this period.

The sale of recently killed wildfowl was prohibited after March 15th by the Act of 1880. The present Act alters this date to February 28th. This date could not be fixed earlier because a week must be allowed for the sale of birds which lawfully may be killed up to February 20th. The sale of wild duck and geese imported at a time of year when importation is legal is not prohibited. This is a provision in the Act which, in the opinion of some who would wish to see all sale during the "close time" prohibited, is not adequate. As, however, by the aid of cold-storage, birds are brought great distances from countries where they have been legally shot during the open season, and can be kept in cold-storage for many months, the main objects of the measure are not helped by prohibiting the sale of such birds.

It cannot be denied that the restrictions with regard to sale in Great Britain are not so comprehensive as in the U.S.A. where all sale and traffic in wildfowl is prohibited, but at the same time in her case this provision is not so vital. Nevertheless, by prohibiting all import during the breeding season, the demand for freshly killed duck (which are of far greater value and in much greater demand than frozen birds) from neighbouring countries, where thousands are caught annually in decoys during this period, is cut off and a great step has therefore been made.

The Act is admittedly not perfect. There were many members of the Wildfowl Committee who wished the "close time" to extend to September 1st or even later, but it is a solid contribution to the maintenance of the stock of wildfowl in Europe. The Act gives five and a half months complete protection to wildfowl (except where application is made for a shortened period on the coast) with the possibility of this time being extended in any area.

The terms of the 1939 Act are as follows:

WILD BIRDS (DUCK AND GEESE) PROTECTION ACT 1939

An Act to provide for the further protection of wild duck and wild geese. [25th May 1939.]

BE it enacted by the King's most Excellent Majesty, by and with the advice and consent of the Lords Spiritual and Temporal, and Commons, in this present Parliament assembled, and by the authority of the same, as follows:—

1.—(1) The period in any year during which, under section three of the principal Act, it is unlawful—

(*a*) to shoot or attempt to shoot, or use a boat for the purpose of shooting or causing to be shot, any wild duck or wild goose; or

(*b*) to use any lime, trap, snare, net or other instrument for the purpose of taking any wild duck or wild goose;

shall, instead of beginning with the second day of March and ending with the thirty-first day of July, begin with the first day of February and end with the eleventh day of August.

(2) The date in any year after which, under the said section three, it is unlawful during the said period to expose or offer for sale, or have control or possession of, any wild duck or wild goose recently killed or taken shall be the twenty-eighth day of February instead of the fifteenth day of March.

(3) After the commencement of this Act—

(*a*) no order shall be made under the power conferred by section eight of the principal Act making any provision whereby the time during which the killing and taking of wild duck or wild geese or any kind thereof is prohibited is to begin with any date later than the first day of February or end with any date earlier than the eleventh day of August, and

(*b*) no order shall be made under the power conferred by section nine of the principal Act exempting any county or any part or parts thereof from the operation of that Act as respects wild duck or wild geese or any kind thereof,

and any order made under the said sections before the commencement of this Act shall cease to have effect if and in so far as it makes any such provision or exemption as aforesaid:

Provided that nothing in this subsection shall prohibit the making after the commencement of this Act of any such order as is mentioned in the next following subsection.

(4) The power of the Secretary of State under section eight of the principal Act shall include power to make an order, in pursuance of an application made under the said section after the commencement of this Act, providing that the time during which the killing and taking of wild duck and wild geese is prohibited shall begin with such date not later than the twenty-first day of February as may be specified in the order in such parts of the county or county borough to which the application relates, being parts contiguous to the low-water mark of ordinary tides, as may be so specified.

2.—(1) Subject to the provisions of this section, the principal Act shall have effect as if wild geese were included in the Schedule thereto (which contains a list of birds in the case of which owners and occupiers of land are not exempted from the provisions of section three of that Act).

(2) A Secretary of State may, on the representation of the council of any county, order that the foregoing subsection shall not extend to that county or any part or parts thereof.

(3) Sections four and six of the Wild Birds Protection Act, 1894 (which relate to the publication of orders and the expenses of county councils) shall apply to any order made under this section as they apply to any order made under that Act.

3.—(1) It shall not be lawful for any person to import into the United Kingdom any dead wild duck or dead wild goose during the period beginning with the first day of February and ending with the eleventh day of August in any year.

(2) During the period aforesaid in every year goods prohibited to be imported by virtue of this section shall be deemed to be included among the goods enumerated and described in the table of prohibitions and restrictions inwards contained in section forty-two of the Customs Consolidation Act, 1876, and the provisions of that Act and of any Act amending or extending that Act shall apply accordingly.

(3) Paragraph (2) of section one of the Wild Birds Protection Act, 1881 (which provides that a person shall not be liable to be convicted under section three of the principal Act of exposing or offering for sale, or having the control or possession of, any wild bird recently killed, if he satisfies the court that the bird was killed in some place to which that Act does not extend), shall not apply in respect of any wild duck or wild goose; but a person shall not be liable to be convicted as aforesaid in respect of any wild duck or wild goose if he satisfies the court before whom he

is charged that the wild duck or wild goose was imported at a time when the importation thereof was not prohibited by this section.

4.—In this Act the following expressions have the meanings hereby respectively assigned to them, that is to say:—

(*a*) "the principal Act" means the Wild Birds Protection Act, 1880, as amended by any subsequent enactment;

(*b*) "wild duck" means eider duck, mallard, pochard, shell-drake, shoveller, smew, teal, wigeon, and wild duck of any other species whatsoever, except merganser and goosander;

(*c*) "wild goose" means a wild goose of any species whatsoever.

5.—(1) This Act may be cited as the Wild Birds (Duck and Geese) Protection Act, 1939, and may be cited together with the Wild Birds Protection Acts, 1880 to 1908, as the Wild Birds Protection Acts, 1880 to 1939.

(2) This Act shall come into operation on the first day of August nineteen hundred and thirty-nine.

(3) Sections one and two and subsection (3) of section three of this Act shall not extend to Northern Ireland.

APPEAL TO THE COUNTRIES OF EUROPE

If all countries of Europe would accord an adequate "close season", the future of the common stock of wild duck and geese would be more assured. It is also of great importance that all countries should co-operate in prohibiting the import of wildfowl during the "close season" in order to assist the endeavours of their neighbours to secure the preservation of wild geese and duck in their respective countries.

CLOSE TIMES OF WILDFOWL

[a] Completely protected during the whole year in the provinces of Jämtland, Västernoirland and certain other districts.
[b] During hard winters the Ministry of Agriculture may order the Close Season to be shortened.
[c] Completely protected during the whole year in the provinces of Bohus and Göteborg.
[d] Along the coast the males may be killed up to April 30.
[e] Protected during the whole year in several localities.
[f] Young are also protected. In Petsamo municipality the Eiders (♂, ♀ and young) may be killed from Sept. 1 to Feb. 28.
[g] In decoys with nets killing is allowed from Oct. 15 to Jan. 31.
[h] May be killed at any time when flying in skeins.

Where more than one period is given, this includes the variation of Close Time in different districts.

SPECIES	AUSTRIA	BELGIUM	BULGARIA	CZECHOSLOVAKIA
Greylag Goose *Anser anser anser*	Feb. 28–July 1 Apr. 15–Aug. 2	[g] Apr. 15–July 15	Apr. 14–Aug. 15	—
Bean Goose *Anser fabalis fabalis*	do	do	do	—
Pinkfooted Goose *Anser fabalis brachyrhynchus*	do	do	do	—
Brent Goose *Branta bernicla*	do	do	do	—
Shelduck *Tadorna tadorna* (L.)	do	do	do	—
Mallard *Anas platyrhyncha platyrhyncha*	do	[g] Feb. 28–July 15	do	Dec. 31–June 1
Gadwall *Anas strepera* L.	do	[g] Apr. 15–July 15	do	do
Teal *Anas crecca crecca* L.	do	do	do	do
Garganey Teal *Anas querquedula* L.	do	do	do	do
Wigeon *Anas penelope* L.	do	do	do	do
Pintail *Anas acuta acuta* L.	do	do	do	do
Shoveler *Spatula clypeata* (L.)	do	do	do	do
Red-crested Pochard *Netta rufina* (Pall.)	do	do	do	do
Common Pochard *Aythya ferina* (L.)	do	do	do	do
Ferruginous Duck *Aythya nyroca nyroca* (Güld.)	do	do	do	do
Tufted Duck *Aythya fuligula* (L.)	do	do	do	do
Scaup Duck *Aythya marila marila*	do	do	do	do
Golden-eye *Bucephala clangula clangula* (L.)	do	do	do	do
Longtailed Duck *Clangula hyemalis* (L.)	do	do	do	do
Eider ♂ *Somateria mollissima mollissima* (L.)	do	do	do	do
Eider ♀ *Somateria mollissima mollissima* (L.)	do	do	do	do
Common Scoter *Melanitta nigra nigra* (L.)	do	do	do	do
Velvet Scoter *Melanitta fusca fusca* (L.)	do	do	do	do
Goosander *Mergus merganser merganser* L.	No close time	do	do	do
Red-breasted Merganser *Mergus serrator* L.	do	do	do	do
Smew *Mergus albellus* L.	do	do	do	do

DENMARK	ICELAND	FAROE ISLANDS	ESTHONIA	*SPECIES*
Dec. 31–Aug. 1	Mar. 31–Aug. 1	All the year	Nov. 30–Aug. 16	Greylag Goose *Anser anser anser*
No close time	—	—	May 14–Aug. 16	Bean Goose *Anser fabalis fabalis*
do	Mar. 31–Aug. 1	—	do	Pinkfooted Goose *Anser fabalis brachyrhynchus*
do	No close time	Jan. 31–Nov. 1	May 31–Aug. 16	Brent Goose *Branta bernicla*
All the year	Mar. 31–Aug. 1	—	Dec. 31–July 26	Shelduck *Tadorna tadorna* (L.)
Dec. 31–Aug. 1	do	Feb. 28–Sept. 1	Nov. 30–July 26	Mallard *Anas platyrhyncha platyrhyncha*
do	do	No close time	do	Gadwall *Anas strepera* L.
do	do	do	do	Teal *Anas crecca crecca* L.
do	do	do	do	Garganey Teal *Anas querquedula* L.
do	do	do	May 14–July 26	Wigeon *Anas penelope* L.
do	do	do	Nov. 30–July 26	Pintail *Anas acuta acuta* L.
do	—	—	Dec. 31–July 26	Shoveler *Spatula clypeata* (L.)
—	—	—	—	Red-crested Pochard *Netta rufina* (Pall.)
Jan. 31–Aug. 1	Mar. 31–Aug. 1	No close time	May 14–July 26	Common Pochard *Aythya ferina* (L.)
[b] Feb. 28–Sept. 1	do	do	do	Ferruginous Duck *Aythya nyroca nyroca* (Güld.)
do	do	do	do	Tufted Duck *Aythya fuligula* (L.)
do	do	do	do	Scaup Duck *Aythya marila marila*
do	do	do	do	Golden-eye *Bucephala clangula clangula* (L.)
do	do	do	do	Longtailed Duck *Clangula hyemalis* (L.)
All the year	All the year	All the year	do	Eider ♂ *Somateria mollissima mollissima* (L.)
do	do	do	Dec. 14–July 26	Eider ♀ *Somateria mollissima mollissima* (L.)
[b] Feb. 28–Sept. 1	Mar. 31–Aug. 1	No close time	May 14–July 26	Common Scoter *Melanitta nigra nigra* (L.)
do	do	do	do	Velvet Scoter *Melanitta fusca fusca* (L.)
Jan. 31–Sept. 18	No close time	Feb. 28–Sept. 1	do	Goosander *Mergus merganser merganser* L.
do	do	do	do	Red-breasted Merganser *Mergus serrator* L.
do	—	—	do	Smew *Mergus albellus* L.

SPECIES	FINLAND	FRANCE	GERMANY	GREECE
Greylag Goose *Anser anser anser*	Feb. 28–Aug. 20	Mar. 31–July 18	Mar. 31–July 16	Mar. 1–Aug. 16
Bean Goose *Anser fabalis fabalis*	do	do	do	do
Pinkfooted Goose *Anser fabalis brachyrhynchus*	do	do	do	do
Brent Goose *Branta bernicla*	do	do	do	do
Shelduck *Tadorna tadorna* (L.)	All the year	do	All the year	do
Mallard *Anas platyrhyncha platyrhyncha*	Feb. 28–Aug. 20	Mar. 1–July 18	Dec. 31–July 16 Jan. 31–July 16	do
Gadwall *Anas strepera* L.	do	Mar. 31–July 18	do	do
Teal *Anas crecca crecca* L.	do	do	do	do
Garganey Teal *Anas querquedula* L.	do	do	do	do
Wigeon *Anas penelope* L.	do	do	do	do
Pintail *Anas acuta acuta* L.	do	do	do	do
Shoveler *Spatula clypeata* (L.)	do	do	do	do
Red-crested Pochard *Netta rufina* (Pall.)	—	do	All the year	do
Common Pochard *Aythya ferina* (L.)	[d] Feb. 28–Sept. 1	do	Dec. 31–July 16 Jan. 31–July 16	do
Ferruginous Duck *Aythya nyroca nyroca* (Güld.)	—	do	do	do
Tufted Duck *Aythya fuligula* (L.)	[d] Feb. 28–Sept. 1	do	do	do
Scaup Duck *Aythya marila marila*	do	do	do	do
Golden-eye *Bucephala clangula clangula* (L.)	[d] Feb. 28–Aug. 20	do	do	do
Longtailed Duck *Clangula hyemalis* (L.)	Feb. 28–Sept. 1 May 20–Sept. 1	do	do	do
Eider ♂ *Somateria mollissima mollissima* (L.)	Feb. 28–June 1	do	All the year	do
Eider ♀ *Somateria mollissima mollissima* (L.)	[f] All the year	do	do	do
Common Scoter *Melanitta nigra nigra* (L.)	Feb. 28–Sept. 1 May 20–Sept. 1	do	Dec. 31–July 16 Jan. 31–July 16	do
Velvet Scoter *Melanitta fusca fusca* (L.)	Feb. 28–Sept. 1	do	do	do
Goosander *Mergus merganser merganser* L.	[d] Feb. 28–Sept. 1	do	Mar. 31–Sept. 1	do
Red-breasted Merganser *Mergus serrator* L.	do	do	do	do
Smew *Mergus albellus* L.	All the year	do	do	do

HOLLAND	HUNGARY	ITALY	LATVIA	SPECIES
Apr. 9–Aug. 2	—	Mar. 31–Aug. 10 Apr. 10–Aug. 10	May 14–July 21	Greylag Goose *Anser anser anser*
do	—	do	do	Bean Goose *Anser fabalis fabalis*
do	—	do	do	Pinkfooted Goose *Anser fabalis brachyrhynchus*
do	Apr. 15–July 1	do	do	Brent Goose *Branta bernicla*
All the year	do	do	Dec. 14–July 21	Shelduck *Tadorna tadorna* (L.)
Feb. 5–July 26 Feb. 26–Aug. 2 Apr. 5–Aug. 27	[h] Feb. 28–July 1	do	do	Mallard *Anas platyrhyncha platyrhyncha*
do	[h] Apr. 15–July 1	do	do	Gadwall *Anas strepera* L.
do	do	do	do	Teal *Anas crecca crecca* L.
do	do	do	do	Garganey Teal *Anas querquedula* L.
do	do	do	do	Wigeon *Anas penelope* L.
do	do	do	do	Pintail *Anas acuta acuta* L.
do	do	do	do	Shoveler *Spatula clypeata* (L.)
do	do	do	do	Red-crested Pochard *Netta rufina* (Pall.)
do	do	do	do	Common Pochard *Aythya ferina* (L.)
do	do	do	do	Ferruginous Duck *Aythya nyroca nyroca* (Güld.)
do	do	do	do	Tufted Duck *Aythya fuligula* (L.)
do	do	do	do	Scaup Duck *Aythya marila marila*
do	do	do	do	Golden-eye *Bucephala clangula clangula* (L.)
do	do	do	do	Longtailed Duck *Clangula hyemalis* (L.)
All the year	Apr. 15–July 1	do	do	Eider ♂ *Somateria mollissima mollissima* (L.)
do	do	do	do	Eider ♀ *Somateria mollissima mollissima* (L.)
Feb. 5–July 26 Feb. 26–Aug. 2 Apr. 5–Aug. 27	[h] Apr. 15–July 1	do	do	Common Scoter *Melanitta nigra nigra* (L.)
do	do	do	do	Velvet Scoter *Melanitta fusca fusca* (L.)
All the year	Apr. 15–July 1	do	do	Goosander *Mergus merganser merganser* L.
do	do	do	do	Red-breasted Merganser *Mergus serrator* L.
do	do	do	do	Smew *Mergus albellus* L.

SPECIES	LITHUANIA	LUXEMBOURG	NORWAY	SVALBARD
Greylag Goose *Anser anser anser*	Feb. 14–July 16	Apr. 25–Aug. 22	Mar. 14–Aug. 15	—
Bean Goose *Anser fabalis fabalis*	do	do	do	—
Pinkfooted Goose *Anser fabalis brachyrhynchus*	do	do	do	June 15–Aug. 16
Brent Goose *Branta bernicla*	do	do	do	do
Shelduck *Tadorna tadorna* (L.)	do	Mar. 31–Aug. 22	do	—
Mallard *Anas platyrhyncha platyrhyncha*	♂ Apr. 14–July 16 ♀ Feb. 14–July 16	do	do	—
Gadwall *Anas strepera* L.	do	do	do	—
Teal *Anas crecca crecca* L.	do	do	do	—
Garganey Teal *Anas querquedula* L.	do	do	do	—
Wigeon *Anas penelope* L.	do	do	do	—
Pintail *Anas acuta acuta* L.	do	do	do	No close time
Shoveler *Spatula clypeata* (L.)	do	do	do	—
Red-crested Pochard *Netta rufina* (Pall.)	do	Apr. 25–Aug. 22	do	—
Common Pochard *Aythya ferina* (L.)	do	do	do	—
Ferruginous Duck *Aythya nyroca nyroca* (Güld.)	do	do	—	—
Tufted Duck *Aythya fuligula* (L.)	do	do	Mar. 14–Aug. 15	—
Scaup Duck *Aythya marila marila*	do	do	do	No close time
Golden-eye *Bucephala clangula clangula* (L.)	♂ Apr. 14–July 16 ♀ Feb. 14–July 16	do	do	—
Longtailed Duck *Clangula hyemalis* (L.)	do	do	do	No close time
Eider ♂ *Somateria mollissima mollissima* (L.)	do	do	[e] Mar. 15–Oct. 15	June 15–Aug. 16
Eider ♀ *Somateria mollissima mollissima* (L.)	do	do	[e] do	do
Common Scoter *Melanitta nigra nigra* (L.)	do	do	Mar. 14–Aug. 15	No close time
Velvet Scoter *Melanitta fusca fusca* (L.)	do	do	do	—
Goosander *Mergus merganser merganser* L.	Feb. 14–July 16	do	do	—
Red-breasted Merganser *Mergus serrator* L.	do	do	do	—
Smew *Mergus albellus* L.	do	do	do	—

Poland	Portugal	Rumania	Spain	*SPECIES*
May 14–Aug. 1	—	Mar. 31–July 15	Feb. 15–Aug. 31	Greylag Goose *Anser anser anser*
do	—	do	do	Bean Goose *Anser fabalis fabalis*
do	—	do	do	Pinkfooted Goose *Anser fabalis brachyrhynchus*
do	—	do	do	Brent Goose *Branta bernicla*
—	—	do	do	Shelduck *Tadorna tadorna* (L.)
♂ May 31–July 16 ♀ and young Dec. 31–July 16	—	do	do	Mallard *Anas platyrhyncha platyrhyncha*
do	—	do	do	Gadwall *Anas strepera* L.
do	—	do	do	Teal *Anas crecca crecca* L.
do	—	do	do	Garganey Teal *Anas querquedula* L.
do	—	do	do	Wigeon *Anas penelope* L.
do	—	do	do	Pintail *Anas acuta acuta* L.
do	—	do	do	Shoveler *Spatula clypeata* (L.)
do	—	do	do	Red-crested Pochard *Netta rufina* (Pall.)
do	—	do	do	Common Pochard *Aythya ferina* (L.)
do	—	do	do	Ferruginous Duck *Aythya nyroca nyroca* (Güld.)
do	—	do	do	Tufted Duck *Aythya fuligula* (L.)
do	—	do	do	Scaup Duck *Aythya marila marila*
♂ May 31–July 16 ♀ Dec. 31–July 16	—	do	do	Golden-eye *Bucephala clangula clangula* (L.)
do	—	do	do	Longtailed Duck *Clangula hyemalis* (L.)
do	—	do	do	Eider ♂ *Somateria mollissima mollissima* (L.)
do	—	do	do	Eider ♀ *Somateria mollissima mollissima* (L.)
do	—	do	do	Common Scoter *Melanitta nigra nigra* (L.)
do	—	do	do	Velvet Scoter *Melanitta fusca fusca* (L.)
Dec. 31–July 16	No close time	Apr. 30–July 15	No close time	Goosander *Mergus merganser merganser* L.
do	—	—	—	Red-breasted Merganser *Mergus serrator* L.
do	No close time	Apr. 30–July 15	No close time	Smew *Mergus albellus* L.

SPECIES	SWEDEN	SWITZERLAND
Greylag Goose *Anser anser anser*	[a]Feb. 28–Sept. 1	Feb. 28–Sept. 1
Bean Goose *Anser fabalis fabalis*	[a]do	do
Pinkfooted Goose *Anser fabalis brachyrhynchus*	[a]do	do
Brent Goose *Branta bernicla*	[a]do	do
Shelduck *Tadorna tadorna* (L.)	[ac]Nov. 15–Aug. 1 Nov. 15–Aug. 11	do
Mallard *Anas platyrhyncha platyrhyncha*	[a]Nov. 15–Aug. 7 Nov. 15–Aug. 11	do
Gadwall *Anas strepera* L.	[a]Nov. 15–Aug. 1 Nov. 15–Aug. 11	do
Teal *Anas crecca crecca* L.	[a]Nov. 15–Aug. 1 Nov. 15–Aug. 11	do
Garganey Teal *Anas querquedula* L.	[a]Nov. 15–Aug. 1 Nov. 15–Aug. 11	do
Wigeon *Anas penelope* L.	[a]Nov. 15–Aug. 1 Nov. 15–Aug. 11	do
Pintail *Anas acuta acuta* L.	[a]Nov. 15–Aug. 1 Nov. 15–Aug. 11	do
Shoveler *Spatula clypeata* (L.)	[a]Nov. 15–Aug. 1 Nov. 15–Aug. 11	do
Red-crested Pochard *Netta rufina* (Pall.)	—	do
Common Pochard *Aythya ferina* (L.)	[a]Apr. 20–Apr. 1 May 10–May 1 Nov. 15–Aug. 11 Dec. 31–Sept. 1 Jan. 31–Dec. 1	do
Ferruginous Duck *Aythya nyroca nyroca* (Güld.)	—	do
Tufted Duck *Aythya fuligula* (L.)	[a]Apr. 20–Apr. 1 May 10–May 1 Nov. 15–Aug. 11 Dec. 31–Sept. 1 Jan. 31–Dec. 1	Feb. 28–Sept. 1
Scaup Duck *Aythya marila marila*	[a]Apr. 20–Apr. 1 May 10–May 1 Nov. 15–Aug. 11 Dec. 31–Sept. 1 Jan. 31–Dec. 1	do
Golden-eye *Bucephala clangula clangula* (L.)	[c]Apr. 20–Apr. 1 May 10–May 1 Nov. 15–Aug. 11 Dec. 31–Sept. 1 Jan. 31–Dec. 1	do
Longtailed Duck *Clangula hyemalis* (L.)	[c]Apr. 20–Apr. 1 May 10–May 1 Nov. 15–Aug. 11 Dec. 31–Sept. 1 Jan. 31–Dec. 1	do
Eider ♂ *Somateria mollissima mollissima* (L.)	[a]Apr. 20–Apr. 1 May 10–May 1 Dec. 31–Sept. 16 Jan. 31–Dec. 1	do
Eider ♀ *Somateria mollissima mollissima* (L.)	[a]Apr. 20–Apr. 1 May 10–May 1 Dec. 31–Sept. 16 Jan. 31–Dec. 1	do
Common Scoter *Melanitta nigra nigra* (L.)	Apr. 20–Apr. 1 May 10–May 1 Nov. 15–Aug. 11 Dec. 31–Sept. 1 Jan. 31–Dec. 1	do
Velvet Scoter *Melanitta fusca fusca* (L.)	[c]Apr. 20–Apr. 1 May 10–May 1 Nov. 15–Aug. 11 Dec. 31–Sept. 1 Jan. 31–Dec. 1	do
Goosander *Mergus merganser merganser* L.	—	do
Red-breasted Merganser *Mergus serrator* L.	—	do
Smew *Mergus albellus* L.	—	do

INDEX

For EU product safety concerns, contact us at Calle de José Abascal, 56–1°, 28003 Madrid, Spain or eugpsr@cambridge.org.

www.ingramcontent.com/pod-product-compliance
Ingram Content Group UK Ltd.
Pitfield, Milton Keynes, MK11 3LW, UK
UKHW050342101225
465764UK00031B/1841
* 9 7 8 1 1 0 7 6 1 7 4 8 3 *